AF559582

An Introduction to Climatology

An Introduction to Climatology

Bimal Dhawan

RANDOM PUBLICATIONS
NEW DELHI (INDIA)

An Introduction to Climatology

ISBN 978-93-5111-415-4

Published in 2017 in India by

RANDOM PUBLICATIONS

4376-A/4B, Gali Murari Lal, Ansari Road
New Delhi-110 002
Phone : +91-11-43580356, +91-11-23289044
e-mail: randomexports@gmail.com, sales@randompublications.com, info@randompublications.com

Reprint 2021

Type Setting by: Friends Media, Delhi-110089
Digitally Printed at : Replika Press Pvt. Ltd.

Preface

Weather is the condition of the atmosphere over a brief period of time. For example, we speak of today's weather or the weather this week. Climate represents the composite of day-to-day weather over a longer period of time. A climatologist attempts to discover and explain the impacts of climate so that society can plan its activities, design its buildings and infrastructure, and anticipate the effects of adverse conditions. Although climate is not weather, it is defined by the same terms, such as temperature, precipitation, wind, and solar radiation. Climatology is approached in a variety of ways. Paleoclimatology seeks to reconstruct past climates by examining records such as ice cores and tree rings (dendroclimatology). Paleotempestology uses these same records to help determine hurricane frequency over millennia. The study of contemporary climates incorporates meteorological data accumulated over many years, such as records of rainfall, temperature and atmospheric composition. Knowledge of the atmosphere and its dynamics is also embodied in models, either statistical or mathematical, which help by integrating different observations and testing how they fit together. Modelling is used for understanding past, present and potential future climates. Historical climatology is the study of climate as related to human history and thus focuses only on the last few thousand years. Climate research is made difficult by the large scale, long time periods, and complex processes which govern climate. Climate is governed by physical laws which can be expressed as differential equations. These equations are coupled and nonlinear, so that approximate solutions are obtained by using numerical methods to create global climate models. Climate is sometimes modelled as a stochastic process but this is generally accepted as an approximation to processes that are otherwise too complicated to analyse.

In contrast to meteorology, which focuses on short term weather systems lasting up to a few weeks, climatology studies the frequency and trends of those systems. It studies the periodicity of weather events over years to millennia, as well as changes in long-term average weather patterns, in relation to atmospheric conditions. Climatologists, those who practice climatology, study both the nature of climates – local, regional or global – and the natural or human-induced factors that cause climates to change. Climatology considers the past and can help predict future climate change. Phenomena of climatological interest include the atmospheric boundary layer, circulation patterns, heat transfer (radiative, convective and latent), interactions between the atmosphere and theoceans and land surface (particularly vegetation, land use and topography), and the chemical and physical composition of the atmosphere. A more complicated way of making a forecast, the analog technique requires remembering a previous weather event which is expected to be mimicked by an upcoming event. What makes it a difficult technique to use is that there is rarely a perfect analog for an event in the future. Some call this type of forecasting pattern recognition, which remains a useful method of observing rainfall over data voids such as oceans with knowledge of how satellite imagery relates to precipitation rates over land, as well as the forecasting of precipitation amounts and distribution in the future. A variation on this theme is used in Medium Range forecasting, which is known as teleconnections, when you use systems in other locations to help pin down the location of another system within the surrounding regime. One method of using teleconnections are by using climate indices such as ENSO-related phenomena.

The book addresses varied facets of this subject. The book is very useful to the people engaged with this subject.

—*Bimal Dhawan*

Contents

1

Introduction

Climatology

Climatology is the study of climate, scientifically defined as weather conditions averaged over a period of time, and is a branch of the atmospheric sciences. Basic knowledge of climate can be used within shorter termweather forecasting using analog techniques such as the El Nino – Southern Oscillation (ENSO), the Madden-Julian Oscillation (MJO), the North Atlantic Oscillation (NAO), the Northern Annualar Mode (NAM), the Arctic oscillation (AO), the Northern Pacific (NP) Index, the Pacific Decadal Oscillation (PDO), and the Interdecadal Pacific Oscillation (IPO).

Climate models are used for a variety of purposes from study of the dynamics of the weather and climate system to projections of future climate.

History

The earliest person to hypothesize climate change may have been the medieval Chinese scientist Shen Kuo (1031–1095). Shen Kuo theorized that climates naturally shifted over an enormous span of time, after observing petrified bamboos found underground near Yanzhou (modern day Yan'an, Shaanxi province), a dry-climate area unsuitable for the growth of bamboo. Early climate researchers include Edmund Halley, who published a map of the trade winds in 1686 after a voyage to the southern hemisphere.

Benjamin Franklin (1706-1790) first mapped the course of the Gulf Stream for use in sending mail from the United States to Europe. Francis

Galton (1822-1911) invented the term *anticyclone*. Helmut Landsberg (1906-1985) fostered the use of statistical analysis in climatology, which led to its evolution into a physical science.

Different Approaches

Climatology is approached in a variety of ways. Paleoclimatology seeks to reconstruct past climates by examining records such as ice cores and tree rings (dendroclimatology). Paleotempestology uses these same records to help determine hurricane frequency over millennia. The study of contemporary climates incorporates meteorological data accumulated over many years, such as records of rainfall, temperatureand atmospheric composition. Knowledge of the atmosphere and its dynamics is also embodied in models, either statistical or mathematical, which help by integrating different observations and testing how they fit together.

Modelling is used for understanding past, present and potential future climates. Historical climatology is the study of climate as related to human history and thus focuses only on the last few thousand years. Climate research is made difficult by the large scale, long time periods, and complex processes which govern climate. Climate is governed by physical laws which can be expressed as differential equations. These equations are coupled and nonlinear, so that approximate solutions are obtained by using numerical methods to create global climate models. Climate is sometimes modelled as a stochastic process but this is generally accepted as an approximation to processes that are otherwise too complicated to analyze.

Indices

El Niño-Southern Oscillation

El Niño/La Niña-Southern Oscillation, or ENSO, is a quasiperiodic climate pattern that occurs across the tropical Pacific Ocean roughly every five years. The *Southern Oscillation* refers to variations in the temperature of the surface of the tropical eastern Pacific Ocean (warming and cooling known as *El Niño* and *La Niña* respectively) and in air surface pressure in the tropical western Pacific. The two variations are coupled: the warm oceanic phase, El Niño, accompanies high air surface pressure in the western Pacific, while the cold phase, La Niña, accompanies low air surface pressure in the western Pacific. Mechanisms that cause the oscillation remain under study. The extremes of this climate pattern's oscillations, El Niño and La Niña, cause extreme weather (such as floods and droughts) in many regions of the world.

Developing countries dependent upon agriculture and fishing, particularly those bordering the Pacific Ocean, are the most affected. In popular usage, the El Niño-Southern Oscillation is often called just "El Niño". El Niño is Spanish for "the little boy" and refers to the Christ child, because periodic warming in the Pacific near South America is usually noticed around Christmas.

Definition

El Niño is defined by prolonged differences in Pacific Ocean surface temperatures when compared with the average value. The accepted definition is a warming or cooling of at least 0.5 °C (0.9 °F) averaged over the east-central tropical Pacific Ocean. Typically, this anomaly happens at irregular intervals of 3–7 years and lasts nine months to two years. The average period length is 5 years. When this warming or cooling occurs for only seven to nine months, it is classified as El Niño/La Niña "conditions"; when it occurs for more than that period, it is classified as El Niño/La Nina "episodes".

The first signs of an El Niño are:

1. Rise in surface pressure over the Indian Ocean, Indonesia, and Australia
2. Fall in air pressure over Tahiti and the rest of the central and eastern Pacific Ocean
3. Trade winds in the south Pacific weaken or head east
4. Warm air rises near Peru, causing rain in the northern Peruvian deserts
5. Warm water spreads from the west Pacific and the Indian Ocean to the east Pacific. It takes the rain with it, causing extensive drought in the western Pacific and rainfall in the normally dry eastern Pacific.

El Niño's warm rush of nutrient-poor tropical water, heated by its eastward passage in the Equatorial Current, replaces the cold, nutrient-rich surface water of the Humboldt Current. When El Niño conditions last for many months, extensive ocean warming and the reduction in Easterly Trade winds limits upwelling of cold nutrient-rich deep water and its economic impact to local fishing for an international market can be serious.

Early Stages and Characteristics of El Niño

Although its causes are still being investigated, El Niño events begin when trade winds, part of the Walker circulation, falter for many months. A series of Kelvin waves—relatively warm subsurface waves of water a few centimetres high and hundreds of kilometres wide—cross the Pacific along the equator and create a pool of warm water near South America, where ocean temperatures are normally cold due to upwelling. The weakening of the winds can also create twin cyclones, another sign of a future El Niño. The Pacific Ocean is a heat reservoir that drives global wind patterns, and the resulting change in its temperature alters weather on a global scale. Rainfall shifts from the western Pacific towards the Americas, while Indonesia and India become drier.

Jacob Bjerknes in 1969 contributed to an understanding of ENSO by suggesting that an anomalously warm spot in the eastern Pacific can weaken the east-west temperature difference, disrupting trade winds that push warm water to the west. The result is increasingly warm water towards the east. Several mechanisms have been proposed through which warmth builds up in equatorial Pacific surface waters, and is then dispersed to lower depths by an El Niño event. The resulting cooler area then has to "recharge" warmth for several years before another event can take place.

While not a direct cause of El Niño, the Madden-Julian Oscillation, or MJO, propagates rainfall anomalies eastward around the global tropics in a cycle of 30–60 days, and may influence the speed of development and intensity of El Niño and La Niña in several ways. For example, westerly flows between MJO-induced areas of low pressure may cause cyclonic circulations north and south of the equator.

When the circulations intensify, the westerly winds within the equatorial Pacific can further increase and shift eastward, playing a role in El Niño development. Madden-Julian activity can also produce eastward-propagating oceanic Kelvin waves, which may in turn be influenced by a developing El Niño, leading to a positive feedback loop.

Southern Oscillation

The Southern Oscillation is the atmospheric component of El Niño. This component is an oscillation in surface air pressure between the tropical eastern and the western Pacific Ocean waters.

The strength of the Southern Oscillation is measured by the *Southern Oscillation Index* (SOI). The SOI is computed from fluctuations in the surface air pressure difference between Tahiti and Darwin, Australia. El Niño episodes are associated with negative values of the SOI, meaning that the pressure difference between Tahiti and

Darwin is relatively small. Low atmospheric pressure tends to occur over warm water and high pressure occurs over cold water, in part because of deep convection over the warm water. El Niño episodes are defined as sustained warming of the central and eastern tropical Pacific Ocean. This results in a decrease in the strength of the Pacific trade winds, and a reduction in rainfall over eastern and northern Australia.

Walker Circulation

During non-El Nino conditions, the Walker circulation is seen at the surface as easterly trade winds that move water and air warmed by the sun towards the west. This also creates ocean upwelling off the coasts of Peru and Ecuador and brings nutrient-rich cold water to the surface, increasing fishing stocks. The western side of the equatorial Pacific is characterized by warm, wet low pressure weather as the collected moisture is dumped in the form of typhoons and thunderstorms. The ocean is some 60 centimetres (24 in) higher in the western Pacific as the result of this motion.

Effects of ENSO's Warm Phase (El Niño)

North America

Winters, during the El Niño effect, are warmer and drier than average in the Northwest, Northmidwest, and Northmideast United States, and therefore those regions experience reduced snowfalls. Meanwhile, significantly wetter winters are present in northwest Mexico and the southwest United States including central and southern California, while both cooler and wetter than average winters in northeast Mexico and the southeast United States (including the Tidewater region of Virginia) occur during the El Niño phase of the oscillation. In Canada, both warmer and drier winters (due to forcing of the Polar Jet further north) over much of the country occur, although less variation from normal is seen in the Maritime Provinces. The following summer is warmer and sometimes drier creating a more active than average forest fire season over Central/Eastern Canada. Some believed that the ice-storm in January 1998, which devastated parts of Southern Ontario and Southern Quebec, was caused or accentuated by El Niño's warming effects. El Niño warmed Vancouver for the 2010 Winter Olympics, such that the area experienced a subtropical-like winter during the games.

Summers, during the El Niño effect, are wetter than average in the Northwest, Northmidwest, Northmideast, and mountain regions of the United States. El Niño is credited with suppressing hurricanes and made the 2009 hurricane season the least active in twelve years. El Niño is also associated with increased wave-caused coastal erosion along the United States Pacific Coast.

There is some evidence that El Niño activity is correlated with incidence of red tides off the Pacific coast of California.

Tropical Cyclones

Most tropical cyclones form on the side of the subtropical ridge closer to the equator, then move poleward past the ridge axis before recurving into the main belt of the Westerlies. When the subtropical ridge position shifts due to El Niño, so will the preferred tropical cyclone tracks. Areas west of Japan and Korea tend to experience much fewer September–November tropical cyclone impacts during El Niño and neutral years. During El Niño years, the break in the subtropical ridge tends to lie near 130°E, which would favour the Japanese archipelago. During El Niño years, Guam's chance of a tropical cyclone impact is one-third of the long term average. The tropical Atlantic ocean experiences depressed activity due to increased vertical wind shear across the region during El Niño years.

Elsewhere

In Africa, East Africa - including Kenya, Tanzania, and the White Nile basin - experiences, in the long rains from March to May, wetter-than-normal conditions. There are also drier than normal conditions from December to February in south-central Africa, mainly in Zambia, Zimbabwe, Mozambique, and Botswana. Direct effects of El Niño resulting in drier conditions occur in parts of Southeast Asia and Northern Australia, increasing bush fires, worsening haze, and decreasing air quality dramatically. Drier-than-normal conditions are also in general observed in Queensland, inland Victoria, inland New South Wales, and eastern Tasmania from June to August.

Many ENSO linkages exist in the high southern latitudes around Antarctica. Specifically, El Niño conditions result in high pressureanomalies over the Amundsen and Bellingshausen seas, causing reduced sea ice and increased poleward heat fluxes in these sectors as well as the Ross Sea. The Weddell Sea, conversely, tends to become colder with more sea ice during El Niño. The exact opposite heating and atmospheric pressure anomalies occur during La Niña.

El Niño's effects on Europe are not entirely clear, but it is not nearly as affected as at least large parts of other continents. There is some evidence that an El Niño may cause a wetter, cloudier winter in

Northern Europe and a milder, drier winter in the Mediterranean Sea region. The El Niño winter of 2006/2007 was unusually mild in Europe, and the Alps recorded very little snow coverage that season.

In most recent times, Singapore experienced the driest February in 2010 since records begins in 1869. With only 6.3 millimetres of rain fell in the month and temperatures hitting as high as 35 degrees Celsius on 26 February. The years 1968 and 2005 had the next driest Februaries, when 8.4 mm of rain fell.

Remote Influence on Tropical Atlantic Ocean

A study of climate records has shown that El Niño events in the equatorial Pacific are generally associated with a warm tropical North Atlantic in the following spring and summer. About half of El Niño events persist sufficiently into the spring months for the Western Hemisphere Warm Pool (WHWP) to become unusually large in summer. Occasionally, El Niño's effect on the Atlantic Walker circulation over South America strengthens the easterly trade winds in the western equatorial Atlantic region. As a result, an unusual cooling may occur in the eastern equatorial Atlantic in spring and summer following El Niño peaks in winter. Cases of El Niño-type events in both oceans simultaneously have been linked to severefamines related to the extended failure of monsoon rains.

ENSO and Global Warming

During the last several decades the number of El Niño events increased, and the number of La Niña events decreased. The question is whether this is a random fluctuation or a normal instance of variation for that phenomenon or the result of global climate changes towards global warming.

The studies of historical data show that the recent El Niño variation is most likely linked to global warming. For example, one of the most recent results is that, even after subtracting the positive influence of decadal variation, shown to be possibly present in the ENSO trend, the amplitude of the ENSO variability in the observed data still increases, by as much as 60% in the last 50 years. It is not certain what exact changes will happen to ENSO in the future: Different models make different predictions. It may be that the observed phenomenon of more frequent and stronger El Niño events occurs only in the initial phase of the global warming, and then (e.g., after the lower layers of the ocean get warmer as well), El Niño will become weaker than it was. It may also be that the stabilizing and destabilizing forces influencing the phenomenon will eventually compensate for each other. More research is needed to provide a better answer to that question, but the current results do not completely exclude the possibility of dramatic changes. The ENSO is considered to be a potential tipping element in Earth's climate.

El Niño "Modoki" and Central-Pacific El Niño

The traditional Niño, also called Eastern Pacific (EP) El Niño, involves temperature anomalies in the Eastern Pacific. However, in the last two decades non-traditional El Niños were observed, in which the usual place of the temperature anomaly (Nino 1 and 2) is not affected, but an anomaly arises in the central Pacific (Nino 3.4). The phenomenon is called Central Pacific (CP) El Niño, "dateline" El Niño (because the anomaly arises near the dateline), or El Niño "Modoki" (Modoki is Japanese for "similar, but different").

The effects of the CP El Niño are different from those of the traditional EP El Niño — e.g., the new El Niño leads to more hurricanes more frequently making landfall in the Atlantic. The recent discovery of El Niño Modoki has some scientists believing it to be linked to global warming. However, Satellite data goes back only to 1979. More research must be done to find the correlation and study past El Niño episodes. The first recorded El Niño that originated in the central Pacific and moved towards the east was in 1986. A joint study by the National Aeronautics and Space Administration and the National Oceanic and Atmospheric Administration concluded thatclimate change may contribute to stronger El Niños. El Niño "Modoki" events occurred in 1991-92, 1994–95, 2002–03, 2004–05, and 2009-10. The strongest such Central Pacific El Niño event known occurred in 2009-2010. There was also a Central Pacific El Niño in 2006-2007.

Health and Social Impacts of El Niño

Extreme weather conditions related to the El Niño cycle correlate with changes in the incidence of epidemic diseases. For example, the El Niño cycle is associated with increased risks of some of the diseases transmitted by mosquitoes, such as malaria, dengue, and Rift Valley fever. Cycles of malaria in India, Venezuela, and Colombia have now been linked to El Niño. Outbreaks of another mosquito-transmitted disease, Australian encephalitis (Murray Valley encephalitis - MVE), occur in temperate south-east Australia after heavy rainfall and flooding, which are associated with La Niña events. A severe outbreak of Rift Valley fever occurred after extreme rainfall in north-eastern Kenya and southern Somalia during the 1997-98 El Niño.

ENSO may be linked to civil conflicts. Scientists at the Earth Institute of Columbia University have analyzed data from 1950 to 2004 and suggest that ENSO may have had a role in 21% of all civil conflicts since 1950, with the risk of annual civil conflict doubling from 3% to 6% in countries affected by ENSO during El Niño years relative to La Niña years.

Cultural History and Pre-historic Information

ENSO conditions have occurred at two- to seven year intervals for at least the past 300 years, but most of them have been weak. There is also evidence for strong El Niño events during the early Holocene epoch 10,000 years ago. El Niño affected pre-Columbian Incas and may have led to the demise of the Moche and other pre-Columbian Peruvian cultures. A recent study suggests that a strong El-Niño effect between 1789–93 caused poor crop yields in Europe, which in turn helped touch off the French Revolution. The extreme weather produced by El Niño in 1876–77 gave rise to the most deadly famines of the 19th century.

An early recorded mention of the term "El Niño" to refer to climate occurs in 1892, when Captain Camilo Carrillo told the Geographical society congress in Lima that Peruvian sailors named the warm northerly current "El Niño" because it was most noticeable around Christmas. The phenomenon had long been of interest because of its effects on the guano industry and other enterprises that depend on biological productivity of the sea.

Charles Todd, in 1893, suggested that droughts in India and Australia tended to occur at the same time; Norman Lockyer noted the same in 1904. An El Niño connection with flooding was reported in 1895 by Pezet and Eguiguren. In 1924, Gilbert Walker (for whom the Walker circulation is named) coined the term "Southern Oscillation". The major 1982–83 El Niño led to an upsurge of interest from the scientific community. The period from 1990–1994 was unusual in that El Niños have rarely occurred in such rapid succession. An especially intense El Niño event in 1998 caused an estimated 16% of the world's reef systems to die. The event temporarily warmed air temperature by 1.5 °C, compared to the usual increase of 0.25 °C associated with El Niño events. Since then, mass coral bleaching has become common worldwide, with all regions having suffered "severe bleaching".

Major ENSO events were recorded in the years 1790–93, 1828, 1876–78, 1891, 1925–26, 1972–73, 1982–83, and 1997–98, with 2009-2010 being one of the strongest ever.

La Niña

La Niña is a coupled ocean-atmosphere phenomenon that is the counterpart of El Niño as part of the broader El Niño-Southern Oscillation climate pattern. During a period of La Niña, the sea surface temperature across the equatorial Eastern Central Pacific Ocean will be lower than normal by 3–5 °C. In the United States, an *episode* of La Niña is defined as a period of at least 5 months of La Niña conditions. The name La Niña originates from Spanish, meaning "the girl," analogous to El Niño meaning "the boy."

La Niña, sometimes informally called "anti-El Niño", is the opposite of El Niño, where the latter corresponds instead to a *higher*sea surface temperature by a deviation of at least 0.5 °C, and its effects are often the reverse of those of El Niño. El Niño is famous due to its potentially catastrophic impact on the weather along both the Chilean, Peruvian, New Zealand, and Australiancoasts, among others. It has extensive effects on the weather in North America, even affecting the Atlantic Hurricane Season. La Niña is often, though not always, preceded by an El Niño.

Effects

The results of La Niña are mostly the opposite of those of El Niño; for example, El Niño would cause a dry period in the Midwestern U.S., while La Niña would typically cause a wet period in that area. La Niña often causes drought conditions in the western Pacific; flooding in northern South America; mild wet summers in northern North America, and drought in the southeastern United States. For India, an El Niño is often a cause for concern because of its adverse impact on the south-west monsoon; this happened in 2009. A La Niña, on the other hand, is often beneficial for the monsoon, especially in the latter half. The La Niña that appeared in the Pacific in 2010 probably helped 2010's south-west monsoon end on a favourable note. But then, it also contributed to the deluge in Australia, which resulted in one of that country's worst natural disasters with large parts of Queensland either under water from floods of unusual proportions or being battered by tropical cyclones, including that of category 5 Tropical Cyclone Yasi. It wreaked similar havoc in south-eastern Brazil and flooding that have affected Sri Lanka.

Recent Occurrences

There was a strong La Niña episode during 1988–1989. La Niña also formed in 1995, and in 1999–2000, followed by neutral periods between 2000 and 2002. The La Niña which developed in mid 2007 and lasted until almost 2009, was a moderate one. NOAA confirmed that a moderate La Niña developed in their November El Niño/Southern Oscillation Diagnostic Discussion, and that it would likely continue into 2008. According to NOAA, "Expected La Niña impacts during November – January include a continuation of above-average precipitation over Indonesia and below-average precipitation over the central equatorial Pacific. For the contiguous United States, potential impacts include above average precipitation in the Northern Rockies, Northern California, and in southern and eastern regions of the Pacific Northwest. Below-average precipitation is expected across the southern tier, particularly in the southwestern and southeastern states.

However, an El Niño returned in May–June 2009 and lasted until April 2010. The effects of El Niño in 2009 were already being seen in the fall of 2009 as the remnants of Tropical Storm Ida strengthened into a powerful coastal storm.

A new La Niña episode developed quite quickly in the eastern and central tropical Pacific in mid-2010, and lasted until early 2011. It intensified again in the mid-2011 and is predicted to last at least until early 2012 This La Niña, combined with record-high ocean temperatures in the north-eastern Indian Ocean, has been a large factor in the 2010–2011 Queensland floods, and the quartet of recent heavy snowstorms in North America starting with the December 2010 North American blizzard. The same La Niña event is also a likely cause of a series of tornadoes of above-average severity that struck the Midwestern and Southern United States in the spring of 2011, and is currently a major factor in the drought conditions persisting in the South Central states including Texas, Oklahoma and Arkansas.

In 2011, on a global scale, La Niña events helped keep the average global temperature below recent trends. As a result, 2011 tied with 1997 for the 11th warmest year on record. It was the second coolest year of the 21st century to date, and tied with the second warmest year of the 20th century. A relatively strong phase of La Niña opened the year, dissipated in the spring before re-emerging in October and lasted through the end of the year. When compared to previous La Niña years, the 2011 global surface temperature was the warmest observed. The 2011 globally-averaged precipitation over land was the second wettest year on record, behind 2010. Precipitation varied greatly across the globe. La Niña contributed to severe drought in the Horn of Africa and to Australia's third wettest year in its 112-year period of record.

Pacific Decadal Oscillation

The Pacific Decadal Oscillation (PDO) is a pattern of Pacific climate variability that shifts phases on at least inter-decadal time scale, usually about 20 to 30 years. The PDO is detected as warm or cool surface waters in the Pacific Ocean, north of 20° N. During a "warm", or "positive", phase, the west Pacific becomes cool and part of the eastern ocean warms; during a "cool" or "negative" phase, the opposite pattern occurs. The Pacific (inter-) decadal oscillation was named by Steven R. Hare, who noticed it while studying salmon production pattern results in 1997. The prevailing hypothesis is that the PDO is caused by a "reddening" of the El Niño-Southern Oscillation (ENSO) combined with stochastic atmospheric forcing.

A PDO signal has been reconstructed to 1661 through tree-ring chronologies in the Baja California area. The interdecadal Pacific oscillation (IPO or ID) display similar sea-surface temperature (SST) and sea-level pressure (SLP) patterns, with a cycle of 15–30 years, but affects both the north and south Pacific. In the tropical Pacific, maximum SST anomalies are found away from the equator. This is quite different from the quasi-decadal oscillation (QDO) with a period of 8-to-12 years and maximum SST anomalies straddling the equator, thus resembling the ENSO.

Mechanisms

Several studies have indicated that the PDO index can be reconstructed as the superimposition of tropical forcing and extra-tropical processes. Thus, unlike ENSO, the PDO is not a single physical mode of ocean variability, but rather the sum of several processes with different dynamical origins. At inter-annual time scales the PDO index is reconstructed as the sum of random and ENSO induced variability in the Aleutian low, on decadal timescales ENSO teleconnections, stochastic atmospheric forcing and changes in the North Pacific oceanic gyre circulation contribute approximately equally, additionally sea surface temperature anomalies have some winter to winter persistence due to the reemergence mechanism.

ENSO teleconnections, the atmospheric bridge: ENSO can influence the global circulation pattern thousands of kilometres away from the equatorial Pacific through the "atmospheric bridge". During El Nino events deep convection and heat transfer to the troposphere is enhanced over the anomalously warm sea surface temperature, this ENSO related tropical forcing generates Rossby waves that propagates poleward and eastward and are subsequently refracted back from the pole to the tropics. The planetary waves forms at preferred locations both in the North and South Pacific Ocean and the teleconnection pattern is established within 2–6 weeks. ENSO driven patterns modify surface temperature, humidity, wind and the distribution of cloud over the North Pacific that alter surface heat, momentum and freshwater fluxes and thus induce sea surface temperature, salinity and mixed layer depth (MLD) anomalies.

The atmospheric bridge is more effective during boreal winter when the deepened Aleutian low results in stronger and cold northwesterly winds over the central Pacific and warm/humid southerly winds along the North American west coast, the associated changes in the surface heat fluxes and to a lesser extent Ekman transport creates negative sea surface temperature anomalies and a deepened MLD in the central pacific and warm the ocean from the Hawaii to the Bering Sea.

SST reemergence: Midlatitude SST anomaly patterns tend to recur from one winter to the next but not during the intervening summer, this process occurs because of the strong mixed layer seasonal cycle. The mixed layer depth over the North Pacific is deeper, typically 100-200m, in winter than it is in summer and thus SST anomalies that forms during winter and extend to the base of the mixed layer are sequestered beneath the shallow summer mixed layer when it reforms in late spring and are effectively insulated from the air-sea heat flux. When the mixed layer deepens again in the following autumn/early winter the anomalies may influence again the surface. This process has been named "reemergence mechanism" by Alexander and Deser and is observed over much of the North Pacific Ocean although is more effective in the west where the winter mixed layer is deeper and the seasonal cycle greater.

Stochastic atmospheric forcing: Long term sea surface temperature variation may be induced by random atmospheric forcings that are integrated and reddened into the ocean mixed layer. The stochastic climate model paradigm was proposed by Frankignoul and Hasselmann, in this model a stochastic forcing represented by the passage of storms alter the ocean mixed layer temperature via surface energy fluxes and Ekman currents and the system is damped due to the enhanced (reduced) heat loss to the atmosphere over the anomalously warm (cold) SST via turbulent energy and longwave radiative fluxes, in the simple case of a linear negative feedback the model can be written as:

$$\frac{dy}{dt} = v(t) - \lambda t$$

where v is the random atmospheric forcing, λ is the damping rate (positive and constant) and y is the response.

The variance spectrum of y is:

$$G(w) = \frac{F}{w^2 + \lambda^2}$$

where F is the variance of the white noise forcing and w is the frequency, an implication of this equation is that at short time scales (w>>λ) the variance of the ocean temperature increase with the square of the period while at longer timescales (w<<l, ~150 months) the damping process dominates and limits sea surface temperature anomalies so that the spectra became white.

Thus an atmospheric white noise generates SST anomalies at much longer timescales but without spectral peaks. Modelling studies suggest that this process contribute to as much as 1/3 of the PDO variability at decadal timescales.

Ocean dynamics: Several dynamic oceanic mechanisms and SST-air feedback may contribute to the observed decadal variability in the North Pacific Ocean. SST variability is stronger in the Kuroshio Oyashio extension (KOE) region and is associated with changes in the KOE axis and strength, that generates decadal and longer time scales SST variance but without the observed magnitude of the spectral peak at ~10 years, and SST-air feedback. Remote reemergence occurs in regions of strong current such as the Kuroshio extension and the anomalies created near the Japan may reemerge the next winter in the central pacific.

- *Advective resonance:* Saravanan and McWilliams have demonstrated that the interaction between spatially coherent atmospheric forcing patterns and an advective ocean shows periodicities at preferred time scales when non-local advective effects dominates over the local sea surface temperature damping. This "advective resonance" mechanism may generate decadal SST variability in the Eastern North Pacific associated with the anomalous Ekman advection and surface heat flux.
- *North Pacific oceanic gyre circulation:* Dynamic gyre adjustments are essential to generate decadal SST peaks in the North Pacific, the process occurs via westward propagating oceanic Rossby waves that are forced by wind anomalies in the central and

eastern Pacific Ocean. The quasigeostrophic equation for long non-dispersive Rossby Waves forced by large scale wind stress can be written as:

$$\frac{\partial h}{\partial t} - c\frac{\partial h}{\partial t} = \frac{-\nabla \times \vec{\tau}}{\rho_0 f_0}$$

where h is the upper-layer thickness anomaly, τ is the wind stress, c is the Rossby wave speed that depends on latitude, $ñ_0$ is the density of sea water and f_0 is the Coriolis parameter at a reference latitude. The response time scale is set by the Rossby waves speed, the location of the wind forcing and the basin width, at the latitude of the Kuroshio Extension c is 2.5 cm s^{-1} and the dynamic gyre adjustement timescale is ~(5)10 years if the Rossby wave was initiated in the (central) eastern Pacific Ocean.

If the wind white forcing is zonally uniform it should generate a red spectrum in which h variance increase with the period and reaches a constant amplitude at lower frequencies without decadal and interdecadal peaks, however low frequencies atmospheric circulation tends to be dominated by fixed spatial patterns so that wind forcing is not zonally uniform, if the wind forcing is zonally sinusoidal then decadal peaks occurs due to resonance of the forced basin-scale Rossby waves.

The propagation of h anomalies in the western pacific changes the KOE axis and strength and impact sst due to the anomalous geostrophic heat transport. Recent studiessuggest that Rossby waves excited by the Aleutian low propagates the PDO signal from the North Pacific to the KOE through changes in the KOE axis while Rossby waves associated with the NPO propagates the North Pacific Gyre oscillation signal through changes in the KOE strength.

Reconstructions and Regime Shifts

The PDO index has been reconstructed using tree rings and other hydrologically sensitive proxies from west North America and Asia.

MacDonald and Case reconstructed the PDO back to 993 using tree rings from California and Alberta. The index shows a 50-70 year periodicity but this is a strong mode of variability only after 1800, a persistent negative phase occurred during medieval times (993-1300) which is consistent with la nina conditions reconstructed in the tropical Pacific and multi-century droughts in the South-West United States. Several regime shifts are apparent both in the reconstructions and instrumental data, during the 20th century regime shifts associated with concurrent changes in SST, SLP, land precipitation and ocean cloud cover occurred in 1924/1925,1945/1946 and 1976/1977:

- 1750: PDO displays an unusually strong oscillation.
- 1924/1925: PDO changed to a "warm" phase.
- 1945/1946: The PDO changed to a "cool" phase, the pattern of this regime shift is similar to the 1970s episode with maximum amplitude in the subarctic and subtropical front but with a greater signature near the Japan while the 1970s shift was stronger near the American west coast.
- 1976/1977: PDO changed to a "warm" phase.
- 1988/1989:A weakening of the Aleutian low with associated SST changes was observed, in contrast to others regime shifts this change appears to be related to concurrent extratropical oscillation in the North Pacific and North Atlantic rather than tropical processes.
- 1997/1998: Several changes in Sea surface temperature and marine ecosystem occurred in the North Pacific after 1997/1998, in contrast to prevailing anomalies observed after the 1970s shift SST declined along the United States west coast and substantial changes in the populations of salmon, anchovy and sardine were observed, however the spatial pattern of the SST change was different with a meridional SST seesaw in the central and western Pacific that resemble a strong shift in the North Pacific Gyre Oscillation rather than the PDO structure, this pattern dominated much of the North Pacific SST variability after 1989.

Predictability

NOAA's forecast use a linear inverse modelling (LIM) method to predict the PDO, LIM assumes that the PDO can be separated into a linear deterministic component and a non-linear component represented by random fluctuations.

Much of the LIM PDO predictability arises from ENSO and the global trend rather than extra-tropical processes and is thus limited to ~4 season, the prediction is consistent with the seasonal footprinting mechanism in which an optimal SST structure evolve into the ENSO mature phase 6–10 months later that subsequently impact the North Pacific Ocean SST via the atmospheric bridge. Skills in predicting decadal PDO variability could arise from taking into account the impact of the externally forced and internally generated pacific variability.

Related Patterns

- ENSO tends to lead PDO/IPO cycling.
- Shifts in the IPO change the location and strength of ENSO activity. The South Pacific Convergence Zone moves northeast during El Niño and southwest during La Niña events. The same movement takes place during positive IPO and negative IPO phases respectively. (Folland et al., 2002)
- Interdecadal temperature variations in China are closely related to those of the NAO and the NPO.
- The amplitudes of the NAO and NPO increased in the 1960s and interannual variation patterns changed from 3–4 years to 8–15 years.
- Sea level rise is affected when large areas of water warm and expand, or cool and contract.

Madden–Julian Oscillation

The Madden–Julian oscillation (MJO) is the largest element of the intraseasonal (30–90 days) variability in the tropical atmosphere. It is a large-scale coupling between atmospheric circulation and tropical deep convection. Rather than being a standing pattern (like ENSO) it is a travelling pattern, propagating eastwards at approximately 4 to 8 m/s, through the atmosphere above the warm parts of the Indian and Pacific oceans. This overall circulation pattern manifests itself in various ways, most clearly as anomalous rainfall. This was discovered by Roland Madden and Paul Julian (again the comparison with ENSO is instructive, since their local effects on Peruvian fisheries were discovered long before the global structure of the pattern was recognized). The MJO is characterized by an eastward progression of large regions of both enhanced and suppressed tropical rainfall, observed mainly over the Indian Ocean and Pacific Ocean. The anomalous rainfall is usually first evident over the western Indian Ocean, and remains evident as it propagates over the very warm ocean waters of the western and central tropical Pacific.

This pattern of tropical rainfall then generally becomes nondescript as it moves over the cooler ocean waters of the eastern Pacific (except over the region of warmer water off the west coast of Central America) but occasionally reappears at low amplitude over the tropical Atlantic and higher amplitude over the Indian Ocean. The wet phase of enhanced convection and precipitation is followed by a dry phase where thunderstorm activity is suppressed. Each cycle lasts approximately 30–60 days. Because of this pattern, The MJO is also known as the 30–60 day oscillation, 30–60 day wave, orintraseasonal oscillation.

Behaviour

There are distinct patterns of lower-level and upper-level atmospheric circulation anomalies which accompany the MJO-related pattern of enhanced or decreased tropical rainfall across the tropics. These circulation features extend around the globe and are not confined to only the eastern hemisphere. The Madden–Julian oscillation moves eastward at between 4 metres per second (8.9 mph) and 8 metres per second (18 mph) across the tropics, crossing the Earth's tropics in 30 to 60 days, with the active phase of the MJO tracked using the degree of outgoing long wave radiation which is measured by infrared-sensing geostationary weather satellites. The lower the amount of outgoing long wave radiation, the stronger the thunderstorm complexes, or convection, is within that region.

Enhanced surface westerly winds occur near the east side of the active convection. Ocean currents, up to 100 metres (330 ft) in depth from the ocean surface, follow in phase with the east-wind component of the surface winds. In advance, or to the east, of the MJO enhanced activity, winds aloft are westerly. In its wake, or to the west of the enhanced rainfall area, winds aloft are easterly. These wind changes aloft are due to the divergence present over the active thunderstorms during the enhanced phase. Its direct influence can be tracked poleward as far as 30 degrees latitude from the equator in both northern and southern hemispheres, propagating outward from its origin near the equator at around 1 degree latitude, or 111 kilometres (69 mi), per day.

Local Effects

Connection to the Monsoon

During the Northern Hemisphere summer season the MJO-related effects on the Indian summer monsoon are well documented. MJO-related effects on the North American summer monsoon also occur, though they are relatively weaker. MJO-related impacts on the North American summer precipitation patterns are strongly linked to meridional (i.e. north–south) adjustments of the precipitation pattern in the eastern tropical Pacific. A strong relationship between the leading mode of intraseasonal variability of the North American Monsoon System, the MJO and the points of origin of tropical cyclones is also present. A period of warming sea surface temperatures are found five to ten days prior to a strengthening of MJO-related precipitation across southern Asia.

A break in the Asian monsoon, normally during the month of July, has been attributed to the Madden–Julian oscillation, after its enhanced phase moves off to the east of the region into the open tropical Pacific ocean.

Influence on Tropical Cyclogenesis

Although tropical cyclones occur throughout the boreal warm season (typically May–November) in both the north Pacific and the north Atlantic basins, in any given year there are periods of enhanced/ suppressed activity within the season. There is evidence that the MJO modulates this activity (particularly for the strongest storms) by providing a large-scale environment that is favourable (or unfavourable) for development. MJO-related descending motion is not favourable for tropical storm development. However, MJO-related ascending motion is a favourable pattern for thunderstorm formation within the tropics, which is quite favourable for tropical storm development. As the MJO progresses eastward, the favoured region for tropical cyclone activity also shifts eastward from the western Pacific to the eastern Pacific and finally to the Atlantic basin.

There is an inverse relationship between tropical cyclone activity in the western north Pacific basin and the north Atlantic basin, however. When one basin is active, the other is normally quiet, and vice versa. The main reason for this appears to be the phase of the MJO, which is normally in opposite modes between the two basins at any given time. While this relationship appears robust, the MJO is one of many factors that contribute to the development of tropical cyclones. For example, sea surface temperatures must be sufficiently warm and vertical wind shear must be sufficiently weak for tropical disturbances to form and persist. However, the MJO also influences these conditions that facilitate or suppress tropical cyclone formation. The MJO is monitored routinely by both the USA National Hurricane Centre and the USA Climate Prediction Centre during the Atlantic hurricane (tropical cyclone) season to aid in anticipating periods of relative activity or inactivity.

Downstream Effects

Link to the El Nino-Southern Oscillation

There is strong year-to-year (interannual) variability in MJO activity, with long periods of strong activity followed by periods in which the oscillation is weak or absent. This interannual variability of the MJO is partly linked to the El Niño-Southern Oscillation (ENSO) cycle.

In the Pacific, strong MJO activity is often observed 6 – 12 months prior to the onset of an El Niñoepisode, but is virtually absent during the maxima of some El Niño episodes, while MJO activity is typically greater during a La Niña episode. Strong events in the Madden–Julian oscillation over a series of months in the western Pacific can speed the development of an El Niño or La Niña but usually do not in themselves lead to the onset of a warm or cold ENSO event. However, observations suggest that the 1982-1983 El Niño developed rapidly during July 1982 in direct response to a Kelvin wave triggered by an MJO event during late May. Further, changes in the structure of the MJO with the seasonal cycle and ENSO might facilitate more substantial impacts of the MJO on ENSO. For example, the surface westerly winds associated with active MJO convection are stronger during advancement towards El Niño and the surface easterly winds associated with the suppressed convective phase are stronger during advancement towards La Nina. Globally, the inter annual variability of the MJO is most determined by atmospheric internal dynamics.

North American Winter Precipitation

The strongest impacts of intraseasonal variability on the United States occur during the winter months over the western U.S. During the winter this region receives the bulk of its annual precipitation. Storms in this region can last for several days or more and are often accompanied by persistent atmospheric circulation features. Of particular concern are the extreme precipitation events which are linked to flooding. There is strong evidence for a linkage between weather and climate in this region from studies that have related the ENSO to regional precipitation variability. In the tropical Pacific, winters with weak-to-moderate cold, or La Nina, episodes or ENSO-neutral conditions are often characterized by enhanced 30–60 day MJO activity. A recent example is the winter of 1996–97, which featured heavy flooding in California and in the Pacific Northwest (estimated damage costs of $2.0–3.0 billion at the time of the event) and a very active MJO. Such winters are also characterized by relatively small sea surface temperature anomalies (SSTA) in the tropical Pacific compared to stronger warm and cold episodes. In these winters there is a stronger linkage between the MJO events and extreme west coast precipitation events.

Pineapple Express Events

The typical scenario linking the pattern of tropical rainfall associated with the MJO to extreme precipitation events in the Pacific Northwest features a progressive (i.e. eastward moving) circulation pattern in the tropics and a retrograding (i.e. westward moving) circulation pattern in the mid latitudes of the North Pacific. Typical wintertime weather anomalies preceding heavy precipitation events in the Pacific Northwest are as follows:

1. *7–10 days prior to the heavy precipitation event:* Heavy tropical rainfall associated with the MJO shifts eastward from the eastern Indian Ocean to the western tropical Pacific. A moisture plume extends northeast ward from the western tropical Pacific towards the general vicinity of the Hawaiian Islands. A strong blocking anticyclone is located in the Gulf of Alaska with a strong polar jet stream around its northern flank.
2. *3–5 days prior to the heavy precipitation event:* Heavy tropical rainfall shifts eastward towards the date line and begins to diminish. The associated moisture plume extends further to the northeast, often traversing the Hawaiian Islands. The strong blocking high weakens and shifts westward. A split in the North Pacific jet stream develops, characterized by an increase in the amplitude and areal extent of the upper tropospheric westerly zonal winds on the southern flank of the block and a decrease on its northern flank. The tropical and extra tropical circulation patterns begin to "phase", allowing a developing mid latitude trough to tap the moisture plume extending from the deep tropics.
3. *The heavy precipitation event:* As the pattern of enhanced tropical rainfall continues to shift further to the east and weaken, the deep tropical moisture plume extends from the subtropical central Pacific into the mid latitude trough now located off the west coast of North America. The jet stream at upper levels extends across the North Pacific with the mean jet position entering North America in the northwestern United States. Deep low pressure located near the Pacific Northwest coast can bring up to several days of heavy rain and possible flooding. These events are often referred to as Pineapple Express events, so named because a significant amount of the deep tropical moisture traverses the Hawaiian Islands on its way towards western North America.

Throughout this evolution, retrogression of the large-scale atmospheric circulation features is observed in the eastern Pacific–North American sector. Many of these events are characterized by the progression of the heaviest precipitation from south to north along the

Pacific Northwest coast over a period of several days to more than one week. However, it is important to differentiate the individual synoptic-scale storms, which generally move west to east, from the overall large-scale pattern which exhibits retrogression.

There is a coherent simultaneous relationship between the longitudinal position of maximum MJO-related rainfall and the location of extreme west coast precipitation events. Extreme events in the Pacific Northwest are accompanied by enhanced precipitation over the western tropical Pacific and the region of Southeast Asia called by meteorologists the Maritime Continent, with suppressed precipitation over the Indian Ocean and the central Pacific. As the region of interest shifts from the Pacific Northwest to California, the region of enhanced tropical precipitation shifts further to the east. For example, extreme rainfall events in southern California are typically accompanied by enhanced precipitation near 170°E. However, it is important to note that the overall linkage between the MJO and extreme west coast precipitation events weakens as the region of interest shifts southward along the west coast of the United States. There is case-to-case variability in the amplitude and longitudinal extent of the MJO-related precipitation, so this should be viewed as a general relationship only.

North Atlantic Oscillation

The North Atlantic oscillation (NAO) is a climatic phenomenon in the North Atlantic Ocean of fluctuations in the difference of atmospheric pressure at sea level between the Icelandic low and the Azores high. Through east-west oscillation motions of the Icelandic low and the Azores high, it controls the strength and direction of westerly winds and storm tracks across the North Atlantic. It is part of the Arctic oscillation, and varies over time with no particular periodicity.

The NAO was discovered in the 1920s by Sir Gilbert Walker. Unlike the El Niño-Southern Oscillation phenomenon in the Pacific Ocean, the NAO is a largely atmospheric mode. It is one of the most important manifestations of climate fluctuations in the North Atlantic and surrounding humid climates. The North Atlantic Oscillation is closely related to the Arctic oscillation (AO) or Northern Annular Mode (NAM), but should not be confused with the Atlantic Multidecadal Oscillation (AMO).

Definition

The NAO has multiple possible definitions. The easiest to understand are those based on measuring the seasonal average air pressure difference between stations, such as:

- Lisbon, Portugal and Stykkisholmur/Reykjavik, Iceland
- Ponta Delgada, Azores and Stykkisholmur/Reykjavik, Iceland
- Azores (1865–2002), Gibraltar (1821–2007), and Reykjavik, Iceland

These definitions all have in common the same northern point (because this is the only station in the region with a long record) in Iceland; and various southern points. All are attempting to capture the same pattern of variation, by choosing stations in the "eye" of the two stable pressure areas: TheAzores high and the Icelandic low.

A more complex definition, only possible with more complete modern records generated by numerical weather prediction, is based on the principalempirical orthogonal function (EOF) of surface pressure. This definition has a high degree of correlation with the station-based definition. This then leads onto a debate as to whether the NAO is distinct from the AO/NAM, and if not, which of the two is to be considered the most physically based expression of atmospheric structure (as opposed to the one that most clearly falls out of mathematical expression).

Description

Westerly winds blowing across the Atlantic bring moist air into Europe. In years when westerlies are strong, summers are cool, winters are mild and rain is frequent. If westerlies are suppressed, the temperature is more extreme in summer and winter leading to heatwaves, deep freezes and reduced rainfall. A permanent low-pressure system over Iceland (the Icelandic Low) and a permanent high-pressure system over the Azores (the Azores High) control the direction and strength of westerly winds into Europe. The relative strengths and positions of these systems vary from year to year and this variation is known as the NAO.

A large difference in the pressure at the two stations (a high index year, denoted NAO+) leads to increased westerlies and, consequently, cool summers and mild and wet winters in Central Europe and its Atlantic facade. In contrast, if the index is low (NAO-), westerlies are suppressed, these areas suffer cold winters and storms track southerly towards the Mediterranean Sea. This brings increased storm activity and rainfall to southern Europe and North Africa. Especially during the months of November to April, the NAO is responsible for much of the variability of weather in the North Atlantic region, affecting wind speed and wind direction changes, changes in temperature and moisture distribution and the intensity, number and track of storms. Although having a less direct influence than for Western Europe, the NAO is also believed to have an impact on the weather over much of eastern North America. During the winter, when the index is high (NAO+), the Icelandic low draws a stronger south-westerly circulation over the eastern half of the North American continent which prevents Arctic air from plunging southward. In combination with the El Niño, this effect can produce significantly warmer winters over the northeastern United States and southeastern Canada. Conversely, when the NAO index is low (NAO-), the eastern seaboard and southestern United States can incur winter cold outbreaks more than the norm with associated snowstorms and sub-freezing conditions into Florida, but the phase of this index has a much smaller effect on winter weather patterns elsewhere on the North American continent.

Effects on North Atlantic Sea Level

Under a positive NAO index (NAO+), regional reduction in atmospheric pressure results in a regional rise in sea level due to the 'inverse barometer effect'. This effect is important to both the interpretation of historic sea level records and predictions of future sea level trends, as mean pressure fluctuations of the order of millibars can lead to sea level fluctuations of the order of centimetres.

North Atlantic Hurricanes

By controlling the position of the Azores high, the NAO also influences the direction of general storm paths for major North Atlantic tropical cyclones: a position of the Azores high farther to the south tends to force storms into the Gulf of Mexico, whereas a northern position allows them to track up the North American Atlantic Coast. As paleotempestological research has shown, few major hurricanes struck the Gulf coast during 3000–1400 BC and again during the most recent millennium. These quiescent intervals were separated by a hyperactive period during 1400 BC and 1000 AD, when the Gulf coast was struck frequently by catastrophic hurricanes and their landfall probabilities increased by 3–5 times.

Ecological Effects

Until recently, the NAO had been in an overall more positive regime since the late 1970s, bringing colder conditions to the North-West Atlantic, which has been linked with the thriving populations of Labrador Sea snow crabs, which have a low temperature optimum. The NAO+ warming of the North Sea reduces survival of cod larvae which are at the upper limits of their temperature tolerance, as does the cooling in the Labrador Sea, where the cod larvae are at their lower temperature limits. Though not the critical factor, the NAO+ peak in the early 1990s may have contributed to the collapse of the Newfoundland cod fishery.

On the East Coast of the United States an NAO+ causes warmer temperatures and increased rainfall, and thus warmer, less saline surface water. This prevents nutrient-rich up welling which has reduced productivity. Georges Bank and the Gulf of Maine are affected by this reduced cod catch.

The strength of the NAO is also a determinant in the population fluctuations of the intensively studied Soay sheep.

Winter of 2009-10 in Europe

The winter of 2009-10 in Europe was unusually cold, especially during December, January and February. It is theorized that this may be due to solar activity. The Met Office reported that the UK, for example, had experienced its coldest winter for 30 years. This coincided with an exceptionally negative phase of the NAO. Analysis published in mid-2010 suggests it was caused by a freak combination of an 'El Nino' event and the rare occurrence of an extremely negative NAO.

However, during the winter of 2010-11 in Northern and Western Europe, the Icelandic low, typically positioned west of Iceland and east of Greenland, appeared regularly to the east of Iceland and so allowed exceptionally cold air into Europe from the Arctic. A strong area of high pressure was initially situated over Greenland, reversing the normal wind pattern in the northwestern Atlantic, creating a blocking pattern driving warm air into northeastern Canada and cold air into Western Europe, as was the case during the previous winter. This occurred during a La Niña season, and is connected to the rare Arctic dipole anomaly.

On the other side of the Atlantic both of these winters were mild, especially 2009-2010, which was the record warmest in Canada. The winter of 2010-2011 was particularly above normal in the northern Arctic regions of the country.

The probability of cold winters with much snow in Central Europe rises when the Arctic is covered by less sea ice in summer. Scientists of the Research Unit Potsdam of the Alfred Wegener Institute for Polar and Marine Research in the Helmholtz Association have decrypted a mechanism in which a shrinking summertime sea ice cover changes the air pressure zones in the Arctic atmosphere and impacts on European winter weather.

If there is a particularly large-scale melt of Arctic sea ice in summer, as observed in recent years, two important effects are intensified. Firstly, the retreat of the light ice surface reveals the darker ocean, causing it to warm up more in summer from the solar radiation (ice-albedo feedback mechanism). Secondly, the diminished ice cover can no longer prevent the heat stored in the ocean being released into the atmosphere (lid effect). As a result of the decreased sea ice cover the air is warmed more greatly than it used to be particularly in autumn and winter because during this period the ocean is warmer than the atmosphere.

The warming of the air near to the ground leads to rising movements and the atmosphere becomes less stable. One of these patterns is the air pressure difference between the Arctic and mid-latitudes: the so-called Arctic oscillation with the Azores highs and Iceland lows known from the weather reports. If this difference is high, a strong westerly wind will result which in winter carries warm and humid Atlantic air masses right down to Europe. In the negative phase when pressure differences are low, cold Arctic air can then easily penetrate southward through Europe without being interrupted by the usual westerlies, as has been the case frequently over the last three winters. Model calculations show that the air pressure difference with decreased sea ice cover in the Arctic summer is weakened in the following winter, enabling Arctic cold to push down to mid-latitudes.

2

Atmosphere of Earth

The atmosphere of Earth is a layer of gases surrounding the planet Earth that is retained by Earth's gravity. The atmosphere protects life on Earth by absorbing ultraviolet solar radiation, warming the surface through heat retention (greenhouse effect), and reducing temperature extremes between day and night (the diurnal temperature variation). Atmospheric stratification describes the structure of the atmosphere, dividing it into distinct layers, each with specific characteristics such as temperature or composition. The atmosphere has a mass of about 5×10 kg, three quarters of which is within about 11 km (6.8 mi; 36,000 ft) of the surface. The atmosphere becomes thinner and thinner with increasing altitude, with no definite boundary between the atmosphere andouter space. An altitude of 120 km (75 mi) is where atmospheric effects become noticeable during atmospheric reentry of spacecraft. The Kármán line, at 100 km (62 mi), also is often regarded as the boundary between atmosphere and outer space. Air is the name given to atmosphere used in breathing and photosynthesis. Dry air contains roughly (by volume) 78.09% nitrogen, 20.95% oxygen, 0.93% argon, 0.039% carbon dioxide, and small amounts of other gases. Air also contains a variable amount of water vapour, on average around 1%. While air content and atmospheric pressure varies at different layers, air suitable for the survival of terrestrial plants and terrestrial animals is currently only known to be found in Earth's troposphere and artificial atmospheres.

Atmospheric Chemistry

Atmospheric chemistry is a branch of atmospheric science in which the chemistry of the Earth's atmosphere and that of other planets is studied. It is a multidisciplinary field of research and draws on environmental chemistry, physics, meteorology, computer modelling, oceanography, geology and volcanology and other disciplines. Research is increasingly connected with other areas of study such as climatology.

The composition and chemistry of the atmosphere is of importance for several reasons, but primarily because of the interactions between the atmosphere and living organisms. The composition of the Earth's atmosphere changes as result of natural processes such as volcano emissions, lightning and bombardment by solar particles from corona. It has also been changed by human activity and some of these changes are harmful to human health, crops and ecosystems. Examples of problems which have been addressed by atmospheric chemistry include acid rain, ozone depletion, photochemical smog, greenhouse gases and global warming. Atmospheric chemists seek to understand the causes of these problems, and by obtaining a theoretical understanding of them, allow possible solutions to be tested and the effects of changes in government policy evaluated.

Atmospheric Composition

The ancient Greeks regarded air as one of the four elements, but the first scientific studies of atmospheric composition began in the 18th century. Chemists such as Joseph Priestley, AntoineLavoisier and Henry Cavendish made the first measurements of the composition of the atmosphere.

In the late 19th and early 20th centuries interest shifted towards trace constituents with very small concentrations. One particularly important discovery for atmospheric chemistry was the discovery of ozone by Christian Friedrich Schönbein in 1840.

In the 20th century atmospheric science moved on from studying the composition of air to a consideration of how the concentrations of trace gases in the atmosphere have changed over time and the chemical processes which create and destroy compounds in the air. Two particularly important examples of this were the explanation by Sydney Chapman and Gordon Dobsonof how the ozone layer is created and maintained, and the explanation of photochemical smog by Arie Jan Haagen-Smit. Further studies on ozone issues led to the 1995 Nobel Prize in Chemistry award shared between Paul Crutzen, Mario Molina and Frank Sherwood Rowland. In the 21st century the focus is now shifting again. Atmospheric chemistry is increasingly studied as one part of the Earth system. Instead of concentrating on atmospheric chemistry in isolation the focus is now on seeing it as one part of a single system with the rest of the atmosphere, biosphere and geosphere. An especially important driver for this is the links between chemistry andclimate such as the effects of changing climate on the recovery of the ozone hole and vice versa but also interaction of the composition of the atmosphere with the oceans and terrestrial ecosystems.

Methodology

Observations, lab measurements and modelling are the three central elements in atmospheric chemistry. Progress in atmospheric chemistry is often driven by the interactions between these components and they form an integrated whole. For example observations may tell us that more of a chemical compound exists than previously thought possible. This will stimulate new modelling and labouratory studies which will increase our scientific understanding to a point where the observations can be explained.

Observation

Observations of atmospheric chemistry are essential to our understanding. Routine observations of chemical composition tell us about changes in atmospheric composition over time. One important example of this is the Keeling Curve - a series of measurements from 1958 to today which show a steady rise in of the concentration of carbon dioxide. Observations of atmospheric chemistry are made in observatories such as that on Mauna Loa and on mobile platforms such as aircraft (e.g. the UK's Facility for Airborne Atmospheric Measurements), ships and balloons. Observations of atmospheric composition are increasingly made by satellites with important instruments such as GOME and MOPITT giving a global picture of air pollution and chemistry. Surface observations have the advantage that they provide long term records at high time resolution but are limited in the vertical and horizontal space they provide observations from. Some surface based instruments e.g. LIDAR can provide concentration profiles of chemical compounds and aerosol but are still restricted in the horizontal region they can cover. Many observations are available on line in Atmospheric Chemistry Observational Databases.

Lab Measurements

Measurements made in the labouratory are essential to our understanding of the sources and sinks of pollutants and naturally occurring compounds. Lab studies tell us which gases react with each other and how fast they react. Measurements of interest include reactions in the gas phase, on surfaces and in water. Also of high importance is photochemistry which quantifies how quickly molecules are split apart by sunlight and what the products are plus thermodynamic data such as Henry's law coefficients.

Modelling

In order to synthesise and test theoretical understanding of atmospheric chemistry, computer models (such as chemical transport models) are used. Numerical models solve the differential equations governing the concentrations of chemicals in the atmosphere. They can be very simple or very complicated. One common trade off in numerical models is between the number of chemical compounds and chemical reactions modelled versus the representation of transport and mixing in the atmosphere. For example, a box model might include hundreds or even thousands of chemical reactions but will only have a very crude representation of mixing in the atmosphere. In contrast, 3D models represent many of the physical processes of the atmosphere but due to constraints on computer resources will have far fewer chemical reactions and compounds. Models can be used to interpret observations, test understanding of chemical reactions and predict future concentrations of chemical compounds in the atmosphere. One important current trend is for atmospheric chemistry modules to become one part of earth system models in which the links between climate, atmospheric composition and the biosphere can be studied. Some models are constructed by automatic code generators (e.g. Autochem or KPP). In this approach a set of constituents are chosen and the automatic code generator will then select the reactions involving those constituents from a set of reaction databases. Once the reactions have been chosen the ordinary differential equations (ODE) that describe their time evolution can be automatically constructed.

Structure of the Atmosphere

Principal layers: In general, air pressure and density decrease in the atmosphere as height increases. However, temperature has a more complicated profile with altitude. Because the general pattern of this profile is constant and recognizable through means such as balloon soundings, temperature provides a useful metric to distinguish between atmospheric layers. In this way, Earth's atmosphere can be divided into five main layers. From highest to lowest, these layers are.

Thermosphere

The thermosphere is the layer of the Earth's atmosphere directly above the mesosphere and directly below the exosphere. Within this layer, ultraviolet radiationcauses ionization. The International Space

Station has a stable orbit within the middle of the thermosphere, between 320 and 380 kilometres (200 and 240 mi). Auroras also occur in the thermosphere. Named from the Greek (*thermos*) meaning heat, the thermosphere begins about 80 kilometres (50 mi) above the Earth. At these high altitudes, the residual atmospheric gases sort into strata according to molecular mass. Thermospheric temperatures increase with altitude due to absorption of highly energetic solar radiation by the small amount of residual oxygen still present. Temperatures are highly dependent on solar activity, and can rise to 1,500 °C(2,730 °F). Radiation causes the atmosphere particles in this layer to become electrically charged, enabling radio waves to bounce off and be received beyond the horizon. In the exosphere, beginning at 500 to 1,000 kilometres (310 to 620 mi) above the Earth's surface, the atmosphere turns into space.

The highly diluted gas in this layer can reach 2,500 °C (4,530 °F) during the day. Even though the temperature is so high, one would not feel warm in the thermosphere, because it is so near vacuum that there is not enough contact with the few atoms of gas to transfer much heat. A normal thermometer would read significantly below 0 °C (32 °F), because the energy lost by thermal radiation would exceed the energy acquired from the atmospheric gas by direct contact. In the anacoustic zone above 160 kilometres (99 mi), the density is so low that molecular interactions are too infrequent to permit the transmission of sound. The dynamics of the lower thermosphere (below approximately 120 kilometres (75 mi)) are dominated by atmospheric tide, which is driven, in part, by the very significant diurnal heating. The atmospheric tide dissipates above this level since molecular concentrations do not support the coherent motion needed for fluid flow.

Mesosphere

The mesosphere is the layer of the Earth's atmosphere that is directly above the stratosphere and directly below the thermosphere. In the mesosphere temperature decreases with increasing height. The upper boundary of the mesosphere is the mesopause, which can be the coldest naturally occurring place on Earth with temperatures below 130 K. The exact upper and lower boundaries of the mesosphere vary with latitude and with season, but the lower boundary of the mesosphere is usually located at heights of about 50 km above the Earth's surface and the mesopause is usually at heights near 100 km, except at middle and high latitudes in summer where it descends to heights of about 85 km. The stratosphere, mesosphere and lowest part of the thermosphere are collectively referred to as the "middle atmosphere", which spans heights from approximately 10 to 100 km. The mesopause, at an altitude of 80–90 km (50–56 mi), separates the mesosphere from the thermosphere—the second-outermost layer of the Earth's atmosphere. This is also around the same altitude as the turbopause, below which different chemical species are well mixed due to turbulent eddies. Above this level the atmosphere becomes non-uniform; the scale heights of different chemical species differ by their molecular masses.

Temperature

Within the mesosphere, temperature increases with increasing altitude. This is due to decreasing solar heating and increasing cooling by CO_2 radiative emission. The top of the mesosphere, called the mesopause, is the coldest part of Earth's atmosphere. Temperatures in the upper mesosphere fall as low as "100 °C(173 K; −148 °F), varying according to latitude and season.

Dynamical Features

The main dynamical features in this region are strong zonal (East-West) winds, atmospheric tides, internal atmospheric gravity waves (commonly called "gravity waves") and planetary waves. Most of these tides and waves are excited in the troposphere and lower stratosphere, and propagate upward to the mesosphere. In the mesosphere, gravity-wave amplitudes can become so large that the waves become unstable and dissipate. This dissipation deposits momentum into the mesosphere and largely drives global circulation.

Noctilucent clouds are located in the mesosphere. The upper mesosphere is also the region of the ionosphere known as the *D layer*. The D layer is only present during the day, when some ionization occurs with nitric oxide being ionized by Lyman series-alpha hydrogen radiation. The ionization is so weak that when night falls, and the source of ionization is removed, the free electron and ion form back into a neutral molecule. A 5 km (3.1 mi) deep sodium layer is located between 80–105 km (50–65 mi). Made of unbound, non-ionized atoms of sodium, the sodium layer radiates weakly to contribute to the airglow.

Uncertainties

The mesosphere lies above the maximum altitude for aircraft and below the minimum altitude for orbital spacecraft. It has only been accessed through the use of sounding rockets. As a result, it is the most poorly understood part of the atmosphere. The presence of red sprites and blue jets (electrical discharges or lightning within the lower mesosphere), noctilucent clouds and density shears within the poorly understood layer are of current scientific interest.

Meteors

Millions of meteors enter the atmosphere, an average of 40 tons per year. Within the mesosphere most melt or vaporize as a result of collisions with the gas particles contained there. This results in a higher concentration of iron and other refractory materials reaching the surface.

Stratosphere

The stratosphere is the second major layer of Earth's atmosphere, just above the troposphere, and below the mesosphere. It is stratifiedin temperature, with warmer layers higher up and cooler layers farther down. This is in contrast to the troposphere near the Earth's surface, which is cooler higher up and warmer farther down. The border of the troposphere and stratosphere, the tropopause, is marked by where this inversion begins, which in terms of atmospheric thermodynamics is the equilibrium level. The stratosphere is situated between about 10 km (6 mi) and 50 km (30 mi) altitude above the surface at moderate latitudes, while at the poles it starts at about 8 km (5 mi) altitude.

Ozone and Temperature

Within this layer, temperature increases as altitude increases; the top of the stratosphere has a temperature of about 270 K (–3°C or 29.6°F), just slightly below the freezing point of water. The stratosphere is layered in temperature because ozone (O_3) here absorbs high energy UVB and UVC energy waves from the Sun and is broken down into atomic oxygen (O) and diatomic oxygen (O_2). Atomic oxygen is found prevalent in the upper stratosphere due to the bombardment of UV light and the destruction of both ozone and diatomic oxygen.

The mid stratosphere has less UV light passing through it, O and O_2 are able to combine, and is where the majority of natural ozone is produced. It is when these two forms of oxygen recombine to form ozone that they release the heat found in the stratosphere. The lower stratosphere receives very low amounts of UVC, thus atomic oxygen is not found here and ozone is not formed (with heat as the byproduct). This vertical stratification, with warmer layers above and cooler layers below, makes the stratosphere dynamically stable: there is no regular convection and associated turbulence in this part of the atmosphere. The top of the stratosphere is called the stratopause, above which the temperature decreases with height.

Methane, (CH_4) while not a direct cause of ozone destruction in the stratosphere, does lead to the formation of compounds that destroy ozone. Monoatomic oxygen (O) in the upper stratosphere reacts with methane (CH_4) to form a hydroxyl radical (OH ·). This hydroxyl radical is then able to interact with non-soluble compounds like chlorofluorocarbons, and UV light break off chlorine radicals (Cl ·). These chlorine radicals break off an oxygen atom from the ozone molecule, creating an oxygen molecule (O_2) and a hypochlorite radical (ClO ·). The hypochlorite radical then reacts with an atomic oxygen creating another oxygen molecule and another chlorine radical, thereby preventing the reaction of a monoatomic oxygen with O_2 to create natural ozone.

Aircraft Flight

Commercial airliners typically cruise at altitudes of 9–12 km (30,000–39,000 ft) in temperate latitudes (in the lower reaches of the stratosphere). This optimizes fuel burn, mostly thanks to the low temperatures encountered near the tropopause and low air density, reducing parasitic drag on the airframe. It also allows them to stay above hard weather (extreme turbulence). Because the temperature in the tropopause and lower stratosphere remains constant (or slightly increases) with increasing altitude, very little convective turbulence occurs at these altitudes. Though most turbulence at this altitude is caused by variations in the jet stream and other local wind shears, areas of significant convective activity (thunderstorms) in the troposphere below may produce convective overshoot. Although a few gliders have achieved great altitudes in the powerful thermals in thunderstorms, this is dangerous. Most high altitude flights by gliders use lee wavesfrom mountain ranges and were used to set the current record of 15,447 m (50,679 ft).

Circulation and Mixing

The stratosphere is a region of intense interactions among radiative, dynamical, and chemical processes, in which the horizontal mixing of gaseous components proceeds much more rapidly than in vertical mixing. An interesting feature of stratospheric circulation is the quasi-biennial oscillation (QBO) in the tropical latitudes, which is driven by gravity waves that are convectively generated in the troposphere. The QBO induces a secondary circulation that is important for the global stratospheric transport of tracers, such asozone or water vapor.

In northern hemispheric winter, sudden stratospheric warmings, caused by the absorption of Rossby waves in the stratosphere, can often be observed.

Life

Bacterial life survives in the stratosphere, making it a part of the biosphere. Also, some bird species have been reported to fly at the lower levels of the stratosphere. On November 29, 1975, a Rüppell's Vulture was reportedly ingested into a jet engine 11,552 m (37,900 ft) above the Ivory Coast, and Bar-headed geese routinely overfly Mount Everest's summit, which is 8,848 m (29,029 ft).

Tropopause

The tropopause is the atmospheric boundary between the troposphere and the stratosphere.

Definition

Going upward from the surface, it is the point where air ceases to cool with height, and becomes almost completely dry. More formally, the tropopause is the region of the atmosphere where the environmental lapse rate changes from positive, as it behaves in the troposphere, to the stratospheric negative one. Following is the exact definition used by the World Meteorological Organization:

The boundary between the troposphere and the stratosphere, where an abrupt change in lapse rate usually occurs. It is defined as the lowest level at which the lapse rate decreases to 2 °C/km or less, provided that the average lapse rate between this level and all higher levels within 2 km does not exceed 2 °C/km. The tropopause as defined above renders as a first-order discontinuity surface, that is, temperature as a function of height is continuous through the tropopause, but the temperature gradient is not.

Location

The troposphere is one of the lowest layers of the Earth's atmosphere; it is located right above the planetary boundary layer, and is the layer in which most weather phenomena take place. The troposphere extends upwards from right above the boundary layer, and ranges in height from an average of 9 km (5.6 mi; 30,000 ft) at the poles, to 17 km (11 mi; 56,000 ft) at the Equator. In the absence of inversions and not considering moisture, the temperature lapse rate for this layer is 6.5°C per kilometre, on average, according to the *U.S. Standard Atmosphere*. A measurement of both the tropospheric and the stratospheric lapse rates helps identifying the location of the tropopause, since temperature increases with height in the stratosphere, and hence the lapse rate becomes negative. The tropopause location coincides with the lowest point at which the lapse rate falls below a prescribed threshold. Given that the tropopause responds to the average temperature of the entire layer that lies underneath it, it is at its peak levels over the Equator, and reaches minimum heights over the poles. On account of this, the coolest layer in the atmosphere lies at about 17 km over the equator. Due to the variation in starting height, the tropopause extremes are referred to as the equatorial tropopause and the polar tropopause.

Given that the lapse rate is not a conservative quantity when the tropopause is considered for stratosphere-troposphere exchanges studies, there exists an alternative definition named *dynamic tropopause.* It is formed with the aid of potential vorticity, which is defined as the product of the isentropic density, i.e. the density that arises from using potential temperature as the vertical coordinate, and the absolute vorticity, given that this quantity attains quite different values for the troposphere and the stratosphere.

Instead of using the vertical temperature gradient as the defining variable, the dynamic tropopause surface is expressed in *potential vorticity units* (PVU), with the tropopause layer typically lying within the 1.5–2 PVU surface in the Northern Hemisphere, although no universal values exist. Given that the absolute vorticity is positive (negative) in the Northern (Southern) hemisphere, the threshold value should be taken as positive (negative) north (south) of the equator. Theoretically, to define a global tropopause in this way, the two surfaces arising from the positive and negative thresholds need to be matched near the equator using another type of surface such as a constant potential temperature surface. Nevertheless, the dynamic tropopause is useless at equatorial latitudes because the isentropes are almost vertical. It is also possible to define the tropopause in terms of chemical composition. For example, the lower stratosphere has much higher ozone concentrations than the upper troposphere, but much lower water vapor concentrations, so appropriate cutoffs can be used.

Phenomena

The tropopause is not a "hard" boundary. Vigorous thunderstorms, for example, particularly those of tropical origin, will overshoot into the lower stratosphere and undergo a brief (hour-order or less) low-frequency vertical oscillation. Such oscillation sets up a low-frequency atmospheric gravity wave capable of affecting both atmospheric and oceanic currents in the region. Most commercial aircraft are flown in the lower stratosphere, just above the tropopause, where clouds are usually absent, as also are significant weather perturbations.

Atmospheric Flow

The flow of the atmosphere generally moves in a west to east direction. This however can often become interrupted, creating a more north to south or south to north flow. These scenarios are often described in meteorology as zonal or meridional. These terms, however, tend to be used in reference to localised areas of atmosphere (at a synoptic scale)). A fuller explanation of the flow of atmosphere around the Earth as a whole can be found in the three-cell model.

Zonal Flow

A zonal flow regime is the meteorological term meaning that the general flow pattern is west to east along the Earth's latitude lines, with weak shortwaves embedded in the flow. The use of the word "zone" refers to the flow being along the Earth's latitudinal "zones". This pattern can buckle and thus become a meridional flow.

Meridional Flow

When the zonal flow buckles, the atmosphere can flow in a more longitudinal (or meridional) direction, and thus the term "meridional flow" arises. Meridional flow patterns feature strong, amplified troughs and ridges, with more north-south flow in the general pattern than west-to-east flow.

Atmospheric Circulation

Atmospheric circulation is the large-scale movement of air, and the means (together with the smaller ocean circulation) by which thermal energy is distributed on the surface of the Earth. The large-scale structure of the atmospheric circulation varies from year to year, but the basic climatological structure remains fairly constant. Individual weather systems - mid-latitude depressions, or tropical convective cells - occur "randomly", and it is accepted that weather cannot be predicted beyond a fairly short limit: perhaps a month in theory, or (currently) about ten days in practice. Nonetheless, as the climate is the average of these systems and patterns - where and when they tend to occur again and again -, it is stable over longer periods of time.

As a rule, the "cells" of Earth's atmosphere shift polewards in warmer climates (e.g. interglacials compared to glacials), but remain largely constant even due to continental drift; they are, fundamentally, a property of the Earth's size, rotation rate, heating and atmospheric depth, all of which change little. Tectonic uplift can significantly alter major elements of it, however - for example the jet stream -, and plate tectonics shift ocean currents. In the extremely hot climates of the Mesozoic, indications of a third desert belt at the Equator has been found; it was perhaps caused by convection. But even then, the overall latitudinal pattern of Earth's climate was not much different from the one today.

Latitudinal Circulation Features

The wind belts girdling the planet are organised into three cells: the Hadley cell, the Ferrel cell, and the Polar cell. Contrary to the impression given in the simplified diagram, the vast bulk of the vertical motion occurs in the Hadley cell; the explanations of the other two cells are complex. Note that there is one discrete Hadley cell that may split, shift and merge in a complicated process over time. Low and high pressures on earth's surface are balanced by opposite relative pressures in the upper troposphere.

Hadley Cell

The Hadley cell mechanism is well understood. The atmospheric circulation pattern that George Hadley described to provide an explanation for the trade winds matches observations very well. It is a closed circulation loop, which begins at the equator with warm, moist air lifted aloft in equatoriallow pressure areas (the Intertropical Convergence Zone, ITCZ) to the tropopause and carried poleward. At about 30°N/S latitude, it descends in a high pressure area. Some of the descending air travels equatorially along the surface, closing the loop of the Hadley cell and creating the Trade Winds.

Though the Hadley cell is described as lying on the equator, it is more accurate to describe it as following the sun's zenith point, or what is termed the "thermal equator," which undergoes a semiannual north-south migration.

Polar Vortex

A polar vortex (also known as Polar cyclones, polar vortices, Arctic cyclones, sub-polar cyclones, and the circumpolar whirl) is a persistent, large-scale cyclone located near one or both of a planet's geographical poles. On Earth, the polar vortices are located in the middle and upper troposphere and the stratosphere. They surround the polar highs and lie in the wake of the polar front. These cold-core low-pressure areas strengthen in the winter and weaken in the summer. They usually spanning 1,000–2,000 kilometres (620–1,240 miles) in which the air is circulating in a counter-clockwise fashion (in the northern hemisphere). The reason for the rotation is the same as any other cyclone, the Coriolis effect.

One centre lies near Baffin Island and the other over northeast Siberia. In the southern hemisphere, it tends to be located near the edge of the Ross ice shelf near 160 west longitude. When the polar vortex is strong, the Westerlies increase in strength. When the polar cyclone is weak, the general flow pattern across mid-latitudes buckles and significant cold outbreaks occur. Ozone depletion occurs within the polar vortex, particularly over the Southern Hemisphere, which reaches a maximum in the spring.

Duration and Power

Polar cyclones are climatological features which hover near the poles year-round. They are weaker during summer and strongest during winter. When the polar vortex is strong, the Westerlies increase in strength. When the polar cyclone is weak, the general flow pattern across mid-latitudes buckles and significant cold outbreaks occur. Extratropical cyclones which occlude and migrate into higher latitudes create cold-core lows within the polar vortex. Volcanic eruptions in the tropics lead to a stronger polar vortex during the winter for as long as two years afterwards. The strength and position of the cyclone shapes the flow pattern across the hemisphere of its influence. An index which is used in the northern hemisphere to gage its magnitude is the Arctic oscillation. The Antarctic polar vortex is more pronounced and persistent than the Arctic one; this is because the distribution of land masses at high latitudes in the northern hemisphere gives rise to Rossby waves which contribute to the breakdown of the vortex, whereas in the southern hemisphere the vortex remains less disturbed. The breakdown of the polar vortex is an extreme event known as a sudden stratospheric warming, here the vortex completely breaks down and an associated warming of 30-50 degrees Celsius over a few days can occur. The Arctic vortex is elongated in shape, with two centres, one normally located over Baffin Island in Canada and the other over northeast Siberia. In rare events, when the general flow pattern is amplified (or meridional), the vortex can push further south as a result of axis interruption, such as during the Winter 1985 Arctic outbreak.

Ozone Depletion

The chemistry of the Antarctic polar vortex has created severe ozone depletion. The nitric acid in polar stratospheric clouds reacts with CFCsto form chlorine, which catalyzes the photochemical destruction of ozone. Chlorine concentrations build up during the polar winter, and the consequent ozone destruction is greatest when the sunlight returns in spring.

These clouds can only form at temperatures below about -80°C. Since there is greater air exchange between the Arctic and the mid-latitudes, ozone depletion at the north pole is much less severe than at the south. Accordingly, the seasonal reduction of ozone levels over the Arctic is usually characterized as an "ozone dent," whereas the more severe ozone depletion over the Antarctic is considered an "ozone hole." This said, chemical ozone destruction in the 2011 Arctic polar vortex attained, for the first time, a level clearly identifiable as an Arctic "ozone hole".

Ongoing Studies

The Australian and US Federal Governments recently awarded funding for a study into how the polar vortex might influence drought in Australia. Scientists hope that the study will glean valuable insight into why droughts in Southern Australia are getting worse, and whether or not there is a direct link between polar climate activity, and weather patterns elsewhere. "One of the big problems we have in planning for drought has to do with understanding whether the drought that we are in right now is a climate-change signal or part of a natural cycle. If we want to understand that we need to understand where the rain is coming from."

Ferrel cell: The Ferrel cell, theorized by William Ferrel (1817-1891), is a secondary circulation feature, dependent for its existence upon the Hadley cell and the Polar cell. It behaves much as an atmospheric ball bearing between the Hadley cell and the Polar cell, and comes about as a result of the eddy circulations (the high and low pressure areas) of the mid-latitudes. For this reason it is sometimes known as the "zone of mixing." At its southern extent (in the Northern hemisphere), it overrides the Hadley cell, and at its northern extent, it overrides the Polar cell. Just as the Trade Winds can be found below the Hadley cell, the Westerlies can be found beneath the Ferrel cell. Thus, strong high pressure areas which divert the prevailing westerlies, such as a Siberian high (which could be considered an extension of the Arctic high), could be said to override the Ferrel cell, making it discontinuous.

While the Hadley and Polar cells are truly closed loops, the Ferrel cell is not, and the telling point is in the Westerlies, which are more formally known as "the Prevailing Westerlies." While the Trade Winds and the Polar Easterlies have nothing over which to prevail, their parent circulation cells having taken care of any competition they might have to face, the Westerlies are at the mercy of passing weather systems. While upper-level winds are essentially westerly, surface winds can vary sharply and abruptly in direction. A low moving polewards or a high moving equator wards maintains or even accelerates a westerly flow; the local passage of a cold front may change that in a matter of minutes, and frequently does. A strong high moving polewards may bring easterly winds for days.

The base of the Ferrel cell is characterized by the movement of air masses, and the location of these air masses is influenced in part by the location of the jet stream, which acts as a collector for the air carried aloft by surface lows (a look at a weather map will show that surface lows follow the jet stream). The overall movement of surface air is from the 30th latitude to the 60th. However, the upper flow of the Ferrel cell is not well defined. This is in part because it is intermediary between the Hadley and Polar cells, with neither a strong heat source nor a strong cold sink to drive convection and, in part, because of the effects on the upper atmosphere of surface eddies, which act as destabilizing influences.

Longitudinal Circulation Features

While the Hadley, Ferrel, and Polar cells are major factors in global heat transport, they do not act alone. Disparities in temperature also drive a set of longitudinal circulation cells, and the overall atmospheric motion is known as the zonal overturning circulation. Latitudinal circulation is the consequence of the fact that incident solar radiation per unit area is highest at the heat equator, and decreases as the latitude increases, reaching its minimum at the poles. Longitudinal circulation, on the other hand, comes about because water has a higher specific heat capacity than land and thereby absorbs and releases more heat, but the temperature changes less than land. Even at mesoscales (a horizontal range of 5 to several hundred kilometres), this effect is noticeable; it is what brings the sea breeze, air cooled by the water, ashore in the day, and carries the land breeze, air cooled by contact with the ground, out to sea during the night.

On a larger scale, this effect ceases to be diurnal (daily), and instead is seasonal or evendecadal in its effects. Warm air rises over the equatorial, continental, and western Pacific Ocean regions, flows eastward or westward, depending on its location, when it reaches the tropopause, and subsides in the Atlantic and Indian Oceans, and in the eastern Pacific. The Pacific Ocean cell plays a particularly important role in Earth's weather. This entirely ocean-based cell comes about as the result of a marked difference in the surface temperatures of the western and eastern Pacific. Under ordinary circumstances, the western

Pacific waters are warm and the eastern waters are cool. The process begins when strong convective activity over equatorial East Asia and subsiding cool air off South America's west coast creates a wind pattern which pushes Pacific water westward and piles it up in the western Pacific. (Water levels in the western Pacific are about 60 cm higher than in the eastern Pacific, a difference due entirely to the force of moving air.)

Walker Circulation

The Walker circulation, also known as the Walker cell, is a conceptual model of the air flow in the tropics in the lower atmosphere (troposphere). According to this model parcels of air follow a closed circulation in the zonal and vertical directions. This circulation, which is roughly consistent with observations, is caused by differences in heat distribution between ocean and land. It was discovered by Gilbert Walker. In addition to motions in the zonal and vertical direction the tropical atmosphere also has considerable motion in the meridional direction as part of, for example, the Hadley Circulation.

Walker's Methodology

Walker determined that the time scale of a year (used by many studying the atmosphere) was unsuitable because geospatial relationships could be entirely different depending on the season. Thus, Walker broke his temporal analysis into December–February, March–May, June–August, and September–November. Walker then selected a number of "centres of action", which included areas such as the Indian Peninsula. The centres were in the hearts of regions with either permanent or seasonal high and low pressures. He also added points for regions where rainfall, wind or temperature was an important control. He examined the relationships of the summer and winter values of pressure and rainfall, first focusing on summer and winter values, and later extending his work to the spring and autumn. He concludes that variations in temperature are generally governed by variations in pressure and rainfall. It had previously been suggested that sunspots could be the cause of the temperature variations, but Walker argued against this conclusion by showing monthly correlations of sunspots with temperature, winds, cloud cover, and rain that were inconsistent.

Walker made it a point to publish all of his correlation findings, both of relationships found to be important as well as relationships that were found to be unimportant. He did this for the purpose of dissuading researchers from focusing on correlations that did not exist.

Mathematical Basis

The statistical model involved in the analysis of atmospheric data that led to the discovery of the Walker circulation is called an autoregressive (AR) process.

Autocorrelation Function

As background, first consider the autocorrelation function. An autocorrelation function in a measure of the dependence of time series values at one time on the values at another time. Given the time series $x(n), n = 1,2,...N$, the autocorrelation function at lag k is defined as:

$$R_{xx}(k) = \frac{1}{(N-k)} \sum_{i=1}^{N-k} x(i)x(i+k).$$

The value of the autocorrelation function at lag 0 is the power of $x(n)$, or its variance if the mean value of is $x(n)$ zero:

$$R_{xx}(0) = \frac{1}{N} \sum_{i=1}^{N} x(i)^2.$$

Moreover, $\sqrt{R_{xx}(\infty)}$, is the mean value for random processes.

The autocorrelation function may be used to detect deterministic components masked in a random background because autocorrelation functions of deterministic data (like sine wave) persist over all time displacements, while autocorrelation functions of stochastic processes tend to zero for large time displacement (for 0-mean time series).

Autoregressive Model

Next, consider the autoregressive model proposed by Walker. Autoregressive Modelling is mathematical modelling of a time series based on the assumption that each value of the series depends only on a weighted sum of the previous values of the same series plus "noise". If $x(j)$ is the j-th value of the time series, the AR model of order p is given by:

$$x(j) = \sum_{i=1}^{p} a_i x(j-i) + n(j).$$

where $n(j)$ is the noise. The order, p, can be considered as an index of the lag within the time series of which data will be considered for the analysis. The larger the lag, the larger the system of equations to be solved.

The AR coefficients can be estimated from the autocorrelation sequence by solving the Yule-Walker equations.

The generalized matrix version of the AR(p) model is given by the equation

$$X_t = \sum_{i=1}^{p} \varphi_i X_{t-i} + \varepsilon_t.$$

Gidon Eshel provides a useful breakdown of the Yule-Walker equations that discusses their relation to between the least squares approach for fitting an AR(p) model.

Yule-Walker Equations

The AR(p) model is given by the equation

$$X_t = \sum_{i=1}^{p} \varphi_i X_{t-i} + \varepsilon_t.$$

It is based on parameters φ_i where $i = 1,\ldots,p$. There is a direct correspondence between these parameters and the covariance function of the process, and this correspondence can be inverted to determine the parameters from the autocorrelation function (which is itself obtained from the covariances). This is done using the Yule-Walker equations:

$$\gamma_m = \sum_{i=1}^{p} \varphi_i \gamma_{m-i} + \sigma_\varepsilon^2 \delta_m$$

where $m = 0,\ldots,p$, yielding $p+1$ equations. γ_m is the autocovariance function of X, σ_ε is the standard deviation of the input noise process, and δ_m is the Kronecker delta function.

Because the last part of the equation is non-zero only if $m = 0$, the equation is usually solved by representing it as a matrix for $m > 0$, thus getting equation

$$\begin{bmatrix} \gamma_1 \\ \gamma_2 \\ \gamma_3 \\ \vdots \end{bmatrix} = \begin{bmatrix} \gamma_0 & \gamma_{-1} & \gamma_{-2} & \cdots \\ \gamma_1 & \gamma_0 & \gamma_{-1} & \cdots \\ \gamma_2 & \gamma_1 & \gamma_0 & \cdots \\ \vdots & \vdots & \vdots & \ddots \end{bmatrix} \begin{bmatrix} \varphi_1 \\ \varphi_2 \\ \varphi_3 \\ \vdots \end{bmatrix}$$

solving all φ. For $m = 0$ have

$$\gamma_0 = \sum_{i=1}^{p} \varphi_i \gamma_{-i} + \sigma_\varepsilon^2$$

which allows us to solve σ_ε^2.

The above equations (the Yule-Walker equations) provide one route to estimating the parameters of an AR(p) model, by replacing the theoretical covariances with estimated values. One way of specifying the estimated covariances is equivalent to a calculation using least squares regression of values X_t on the p previous values of the same series.

Oceanic Effects

The Walker Circulations of the tropical Indian, Pacific, and Atlantic basins result in westerly surface winds in Northern Summer in the first basin and easterly winds in the second and third basins. As a result the temperature structure of the three oceans display dramatic asymmetries. The equatorial Pacific and Atlantic both have cool surface temperatures in Northern Summer in the east, while cooler surface temperatures prevail only in the western Indian Ocean. And these changes in surface temperature reflect changes in the depth of the thermocline.

Changes in the Walker Circulation with time occur in conjunction with changes in surface temperature. Some of these changes are forced externally, such as the seasonal shift of the Sun into the Northern Hemisphere in summer. Other changes appear to be the result of coupled ocean-atmosphere feedback in which, for example, easterly winds cause the sea surface temperature to fall in the east, enhancing the zonal heat contrast and hence intensifying easterly winds across the basin. These anomalous easterlies induce more equatorial upwelling and raise the thermocline in the east, amplifying the initial cooling by the southerlies. This coupled ocean-atmosphere feedback was originally proposed by Bjerknes. From an oceanographic point of view, the equatorial cold tongue is caused by easterly winds. Were the earth climate symmetric about the equator, cross-equatorial wind would vanish, and the cold tongue would be much weaker and have a very different zonal structure than is observed today. The Walker cell is indirectly related to upwelling off the coasts of Peru and Ecuador. This brings nutrient-rich cold water to the surface, increasing fishing stocks.

Synoptic Scale Observations and Concepts

Forcing

Forcing is a term used by meteorologists to describe the situation where a change or an event in one part of the atmosphere causes a strengthening change in another part of the atmosphere. It is usually used to describe connections between upper, middle or lower levels (such as upper-level divergence causing lower level convergence in cyclone formation), but can sometimes also be used to describe such connections over distance rather than height alone. In some respects, tele-connections could be considered a type of forcing.

Divergence and Convergence

An area of convergence is one in which the total mass of air is increasing with time, resulting in an increase in pressure at locations below the convergence level (recall that atmospheric pressure is just the total weight of air above a given point). Divergence is the opposite of convergence - an area where the total mass of air is decreasing with time, resulting in falling pressure in regions below the area of divergence. Where divergence is occurring in the upper atmosphere, there will be air coming in to try to balance the net loss of mass (this is called the principle of mass conservation), and there is a resulting upward motion (positive vertical velocity). Another way to state this is to say that regions of upper air divergence are conducive to lower level convergence, cyclone formation, and positive vertical velocity. Therefore, identifying regions of upper air divergence is an important step in forecasting the formation of a surface low pressure area.

Other Layers

Within the five principal layers determined by temperature are several layers determined by other properties:

- The ozone layer is contained within the stratosphere. In this layer ozone concentrations are about 2 to 8 parts per million, which is much higher than in the lower atmosphere but still very small compared to the main components of the atmosphere. It is mainly located in the lower portion of the stratosphere from about 15–35 km (9.3–22 mi; 49,000–110,000 ft), though the thickness varies seasonally and geographically. About 90% of the ozone in our atmosphere is contained in the stratosphere.
- The ionosphere, the part of the atmosphere that is ionized by solar radiation, stretches from 50 to 1,000 km (31 to 620 mi; 160,000 to 3,300,000 ft) and typically overlaps both the exosphere and the thermosphere. It forms the inner edge of the magnetosphere. It has practical importance because it influences, for example, radio propagation on the Earth. It is responsible for auroras.
- The homosphere and heterosphere are defined by whether the atmospheric gases are well mixed. In the homosphere the chemical composition of the atmosphere does not depend on molecular weight because the gases are mixed by turbulence. The homosphere includes the troposphere, stratosphere, and

mesosphere. Above the *turbopause* at about 100 km (62 mi; 330,000 ft) (essentially corresponding to the mesopause), the composition varies with altitude. This is because the distance that particles can move without colliding with one another is large compared with the size of motions that cause mixing. This allows the gases to stratify by molecular weight, with the heavier ones such as oxygen and nitrogen present only near the bottom of the heterosphere. The upper part of the heterosphere is composed almost completely of hydrogen, the lightest element.

- The planetary boundary layer is the part of the troposphere that is nearest the Earth's surface and is directly affected by it, mainly through turbulent diffusion. During the day the planetary boundary layer usually is well-mixed, while at night it becomes stably stratified with weak or intermittent mixing. The depth of the planetary boundary layer ranges from as little as about 100 m on clear, calm nights to 3000 m or more during the afternoon in dry regions.

The average temperature of the atmosphere at the surface of Earth is 14 °C (57 °F; 287 K) or 15 °C (59 °F; 288 K), depending on the reference.

3

Atmospheric Temperature

Atmospheric temperature is a measure of temperature at different levels of the Earth's atmosphere. It is governed by many factors, including incoming solar radiation, humidity and altitude. When discussing surface temperature, the annual atmospheric temperature range at any geographical location depends largely upon the type of biome, as measured by the Köppen climate classification.

Temperature Versus Height

In the Earth's atmosphere, temperature varies greatly at different heights relative to the Earth's surface. The coldest temperatures lie near the mesopause, an area approximately 85 km (53 mi) to 100 km (62 mi) above the surface. In contrast, some of the warmest temperatures can be found in the thermosphere, which receives strong ionizing radiation at the level of the Van Allen radiation belt. Temperature varies as one moves vertically upwards from the Earth's Surface.

Global Temperature

The concept of a global temperature is commonly used in climatology, and denotes the average temperature of the Earth based on surface, near-surface or tropospheric measurements. These temperature records and measurements are typically acquired using thesatellite or ground instrumental temperature measurements, then usually compiled using a database or computer model. Long-term global temperatures in paleoclimate are discerned using proxy data.

Atmospheric Temperature Range

Atmospheric temperature range is the numerical difference between the minimum and maximum values of temperature observed in a given location.

A temperature range may refer to a period of time (e.g., in a given day, month, year, century) or to an average (average of all temperature ranges in a period of time). The variation in temperature that occurs from the highs of the day to the cool of nights is called diurnal temperature variation.

The size of ground-level atmospheric temperature ranges depends on several factors, such as:

- The average temperature
- The average humidity
- The regime of winds (intensity, duration, variation, temperature, etc.)
- The proximity to large bodies of water, such as the sea

A location which combines an average temperature of 19 degrees Celsius, 60% average humidity and a temperature range of about 10 degrees Celsius around the average temperature (yearly temperature variation) is considered ideal in terms of comfort for the human species. Most of the places with these characteristics are located in the transition between temperate and tropical climates, approximately around the tropics, particularly in the Southern hemisphere (the tropic of Capricorn).

The figure at left shows an example of monthly temperatures recorded at one of such locations, the city of Campinas, state of São Paulo, Brazil, which lies approximately 60 km north of the Capricorn line (latitude of 22 degrees). Average yearly temperature is 22.4 degrees Celsius, ranging from an average minimum of 12.2 degrees to a maximum of 29.9 degrees. The average temperature range is 11.4 degrees. Variability along the year is small (standard deviation of 2.31 for the maximum monthly average and 4.11 for the minimum). One can easily see in the graph another typical phenomenon of temperature ranges, which is its increase during winter (lower average air temperature). In Campinas, for example, the daily temperature range in July (the coolest month of the year) may vary between typically 10 and 24 degrees Celsius (range of 14), while in January, it may range between 20 and 30 degrees Celsius (range of 10).

The effect of latitude, tropical climate, constant gentle wind and sea-side locations clearly show smaller average temperature ranges, smaller variations of temperature, and a higher average temperature (second graph, taken for the same period as Campinas, at Aracaju, capital of the state of Sergipe, also in Brazil, at a latitude of 10 degrees, nearer to the Equator). Average maximum yearly temperature is 28.7 degrees Celsius and average minimum is 21.9. The average temperature range is 5.7 degrees only. Temperature variation along the year in

Aracaju is very damped (standard deviation of 1.93 for the maximum temperature and 2.72 for the minimum temperature).

Lifted Minimum Temperature

The minimum temperature at night does not occur on the ground but few tens of centimetres above the ground. The lowest temperature layer is called *Ramdas layer* after L. A. Ramdas, who first reported this phenomenon in 1932 based on observations at different screen heights at six meteorological centres across India. The phenomenon is attributed to the interaction of thermal radiation effects on atmospheric aerosols and convection transfer close to the ground.

Bond Event

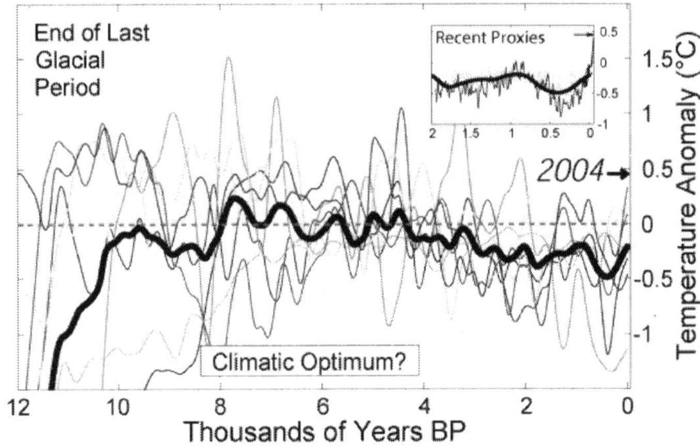

Bond events are North Atlantic climate fluctuations occurring every ≈1,470 ± 500 years throughout the Holocene. Eight such events have been identified, primarily from fluctuations in ice-rafted debris. Bond events may be the interglacial relatives of the glacial Dansgaard–Oeschger events, with a magnitude of perhaps 15–20% of the glacial-interglacial temperature change.

Gerard C. Bond of the Lamont-Doherty Earth Observatory at Columbia University, was the lead author of the paper published in 1997 that postulated the theory of 1,470-year climate cycles in the Holocene, mainly based on petrologic tracers of drift ice in the North Atlantic.

The existence of climatic changes, possibly on a quasi-1,500 year cycle, is well established for the last glacial period from ice cores. Less well established is the continuation of these cycles into the holocene. Bond *et al.* (1997) argue for a cyclicity close to 1470 ± 500 years in the

North Atlantic region, and that their results imply a variation in Holocene climate in this region. In their view, many if not most of the Dansgaard–Oeschger events of the last ice age, conform to a 1,500-year pattern, as do some climate events of later eras, like the Little Ice Age, the 8.2 kiloyear event, and the start of the Younger Dryas.

The North Atlantic ice-rafting events happen to correlate with most weak events of the Asian monsoon over the past 9,000 years, as well as with most aridification events in the Middle East. Also, there is widespread evidence that a H–1,500 yr climate oscillation caused changes in vegetation communities across all of North America.

For reasons that are unclear, the only Holocene Bond event that has a clear temperature signal in the Greenland ice cores is the 8.2 kyr event. The hypothesis holds that the 1,500-year cycle displays nonlinear behaviour and stochastic resonance; not every instance of the pattern is a significant climate event, though some rise to major prominence in environmental history. Causes and determining factors of the cycle are under study; researchers have focused attention on variations in solar output, and "reorganizations of atmospheric circulation." Bond events may also be correlated with the 1800 year lunar tidal cycle.

List of Bond Events

Most Bond events do not have a clear climate signal; some correspond to periods of cooling, others are coincident with aridification in some regions.

- H–1,400 BP (Bond event 1)
- H–2,800 BP (Bond event 2) — correlates with an early 1st millennium BC drought in the Eastern Mediterranean, possibly triggering the collapse of Late Bronze Age cultures.
- H–4,200 BP (Bond event 3) — correlates also with the collapse of the Akkadian Empire and the end of the Egyptian Old Kingdom.
- H–5,900 BP (Bond event 4)
- H–8,100 BP (Bond event 5)
- H–9,400 BP (Bond event 6) — correlates with the Erdalen event of glacier activity in Norway, as well as with a cold event in China.
- H–10,300 BP (Bond event 7)
- H–11,100 BP (Bond event 8) — coincides with the transition from the Younger Dryas to the boreal.

Chilling Requirement

The chilling requirement of a fruit is the minimum period of cold weather after which a fruit-bearing tree will blossom. It is often expressed in chill hours, which can be calculated in different ways, all of which essentially involve adding up the total amount of time in a winter spent at certain temperatures.

Some bulbs have chilling requirements to bloom, and some seeds have chilling requirements to sprout. Biologically, the chilling requirement is a way of ensuring that vernalization occurs.

Chilling Units or Chilling Hours

Chilling unit in agriculture is a metric of a plant's exposure to chilling temperatures. Chilling temperatures extend from freezing point to, depending on the model, 45 °F (7 °C) or even60 °F (16 °C). Stone fruit trees and certain other plants of temperate climate develop next year's buds in the summer. In the autumn the buds go dormant, and the switch to proper, healthy dormancy is triggered by a certain minimum exposure to chilling temperatures. Lack of such exposure results in delayed and substandard foliation, flowering and fruiting. One chilling unit, in the simplest models, is equal to one hour's exposure to the chilling temperature; these units are summed up for a whole season. Advanced models assign different weights to different temperature bands.

Chilling Requirements

According to Fishman, chilling in trees acts in two stages. The first is reversible: chilling helps to build up the precursor to dormancy, but the process can be easily reversed with a rise in temperature. After the level of precursor reaches a certain threshold, dormancy becomes irreversible and will not be affected by short-term warm temperature peaks. Apples have the highest chilling requirements of all fruit trees, followed by apricots and, lastly, peaches. Apple cultivars have a diverse range of permissible minimum chilling: most have been bred for temperate weather, but *Gala* and *Fuji* can be successfully grown in subtropic Bakersfield, California.

Peach cultivars in Texas range in their requirements from 100 chilling units (*FlordaGrande* cultivar, zoned for low chill regions) to 1,000 units (*Surecrop*, zoned for high chill regions). Planting a low-chilling cultivar in a high-chill region risks loss of a year's harvest when an early bloom is hit by a spring frost. A high-chilling cultivar planted in a low-chill region will, quite likely, never fruit at all. A four-year study of *Ruston Red* Alabama peach, which has a threshold of 850 chilling units, demonstrated that a seasonal chilling deficiency of less than 50 units has no effect on harvest. Deficiency of 50 to 100 units may result in loss of up to 50% of expected harvest. Deficiency of 250 hours and more is a sure loss of practically whole harvest; the few fruit will be of very poor quality and have no market value. Rest-breaking agents (e.g. hydrogen cyanamide, trade name *Dormex*), applied in spring, can partially mitigate the effects of insufficient chilling. Dormex can substitute for up to 300 hours of chilling, but an excessive spraying and timing error can easily damage the buds.

Chilling of orange trees has two effects. First, it increases production of carotenoids and decreases chlorophyll content of the fruit, improving their appearance and, ultimately, their market value. Second, the "quasi-dormancy" experienced by orange trees triggers concentrated flowering in spring, as opposed to more or less uniform round-the-year flowering and fruiting in warmer climates. Biennial plants like cabbage, sugar beet, celery and carrots need chilling to develop second-year flowering buds. Excessive chilling in the early stages of a sugar beet seedling, on the contrary, may trigger undesired growth of a flowering stem (bolting) in its first year. This phenomenon has been offset by breeding sugar beet cultivars with a higher minimum chilling threshold. Such cultivars can be seeded earlier than normal without the risk of bolting.

Models

All models require hourly recording of temperatures. The simplest model assigns one chilling unit for every full hour at temperatures below 45 °F (7 °C). A slightly more sophisticated model excludes freezing temperatures, which do not contribute to proper dormancy cycle, and counts only hours with temperatures between 32 °F (0 °C) and 45 °F (7 °C). The Utah model assigns different weight to different temperature bands; a full unit per hour is assigned only to temperatures between 37 °F (3 °C) and 48 °F (9 °C). Maximum effect is achieved at 7 °C (45 °F). Temperatures between 55 °F (13 °C) and 60 °F (16 °C) (the threshold between chilling and warm weather) have zero weight, and higher temperature have negative weights: they reduce the beneficial effects of an already accumulated chilling hours.

Southwick et al. wrote that neither of these models is accurate enough to account for application of rest-breaking agents widely used in modern farming. They advocated the use of a dynamic model taylored to the two-stage explanation of dormancy.

Stone (prunus) Fruit Tree Selection Guidelines to Match Local Weather Conditions

When discussing prunus fruit trees (almonds, apricots, cherries, nectarines, plums, peaches,) there are several climate guidelines to follow for maximum crop yield.

- Select varieties that have a chilling requirement at least 20% less than local averages.
- Selecting a low chill variety in a cold area will result in trees flowering too early and being damaged by late frosts.
- Selecting a high chill variety in warm areas will result in little or no fruit production.
- Early flowering varieties are best in warm climates, late flowering varieties are best in cooler areas.
- Early ripening varieties are best in areas with intense summers, late ripening varieties are best in cooler summers.
- Climate extremes may eliminate certain varieties that would otherwise meet the chilling requirements. For example, the very dry air and intense summer heat as found in Phoenix Arizona may stress a fruit tree beyond its ability to produce quality fruit.
- Terrain can affect the chilling hours received. Open slopes may receive more chilling hours than sheltered areas next to warm buildings.
- Various sellers of fruit trees publish significantly varying chilling hour requirements for the same variety. It is difficult to know the exact requirements. Experiment and ask around for promising local cultivar success stories.

Following the above guidelines here is a practical example. A good apricot for Phoenix Arizona (350 chilling hours) would be Katy apricot with a 200-300 chilling hours requirement. It is early blooming and ripens in May and the tree itself thrives in the intense dry desert heat with adequate regular irrigation. The Katy apricot has no apparent pests or disease problems locally. Planting a Katy apricot only 100 miles north (1000+ chill hours) would likely be fruitless from late frost damage to the flowers. A late ripening apricot variety like Autumn Glo might be a bad choice for Phoenix because the intense long summer heat (115+) might cook the green fruit on the tree and result in strange tastes and other problems with late ripening. The same late ripening variety might also fail in the colder areas 100 mile north because of the shorter summer not allowing enough time to properly ripen before cold weather sets in. A better apricot choice for that colder area might be Goldcot with an 800 chill hour requirement. The late ripening Autumn Glo might be better off in a long cool summer climate along the west coast.

Climate Appraisal

A climate appraisal as can be requested at Climate Appraisal Services, LLC (Climate Appraisal. com), is a unique address-based report for a specific property on climate change (fromglobal warming) and other environmental risks. Information in a climate appraisal report enables property owners and/or buyers to assess for themselves how climate change risks could impact a specific property address in the future, utilizing both historical and projected data from scientific modelling to display those potential risks.

A climate appraisal report includes historical and projected information in the following categories: shoreline reduction from sea level rise, hurricanes, tornadoes, earthquakes, volcanoes, drought, wildfire, flooding, disease, and industrial pollution. A report integrates a wide range of environmental risk information after geocoding an input property address, and combines it with regional commentary from scientists. In the category of sea level rise for coastline states, a report contains a map of an address assuming a 3 foot sea level rise this century to a resolution of 5 miles by 5 miles, and illustrates whether or not it could be submerged from climate change.

Background

Climate Appraisal Services, LLC was launched in November 2006 in a partnership with University of Arizona scientists, with a mission to provide address-based climate reporting to property owners to enable a new channel of climate change communication. Climate Appraisal Services was reviewed in 2007 by USA Today in an article entitled "Website checks your home's climate change risk", as well as other media in 2007 including the Nature journal.

Climate Categories in Viticulture

In viticulture, the climates of wine regions are categorized based on the overall characteristics of the area's climate during thegrowing season. While variations in macroclimate are acknowledged, the climates of most wine regions are categorized (somewhat loosely based on the Köppen climate classification) as being part of a Mediterranean (for example Tuscany), maritime (ex: Bordeaux) or continental climate (ex: Columbia Valley). The majority of the world's premium wine production takes place in one of these three climate categories in locations between the 30th parallel and 50th parallel in both the northern and southern hemisphere. While viticulture does exist in some tropical climates, most notably Brazil, the amount of quality wine production in those areas is so small that the climate effect has not been as extensively studied as other categories.

Influence of Climate on Viticulture

Beyond establishing whether or not viticulture can even be sustained in an area, the climatic influences of a particular area goes a long way in influencing the type of grape varieties grown in a region and the type of viticultural practices that will be used. The presence of adequate sun, heat and water are all vital to the healthy growth and development of grapevines during the growing season. Additionally, continuing research has shed more light on the influence of dormancy that occurs after harvest when the grapevine essentially shuts down and reserves its energy for the beginning of the next year's growing cycle.

In general, grapevines thrive in temperate climates which grant the vines long, warm periods during the crucial flowering, fruit set and ripening periods. The physiological processes of a lot of grapevines begin when temperatures reach around 50 °F (10 °C). Below this temperature, the vines are usually in a period of dormancy. Drastically below this temperature, such as the freezing point of 32 °F (0 °F) the vines can be damaged by frost. When the average daily temperature is between 63 and 68 °F (17 and 20 °C) the vine will begin flowering. When temperatures move into the 80s Fahrenheit (27+ °C) many of the vine's physiological processes are in full stride as grape clusters begin to ripen on the vine. One of the characteristics that differentiates the various climate categories from one another is the occurrence and length of time that these optimal temperatures appear during the growing season.

In addition to temperature, the amount of rainfall (and the need for supplemental irrigation) is another defining characteristics. On average, a grapevine needs around 28 inches (700 mm) of water for sustenance during the growing season, not all of which may be provided by natural rain fall. In Mediterranean and many continental climates, the climate during the growing season may be quite dry and require additional irrigation. In contrast, maritime climates often suffer the opposite extreme of having too much rainfall during the growing season which poses its own viticultural hazards.

Other climate factors such as wind, humidity, atmospheric pressure, sunlight as well as diurnal temperature variations which can define different climate categories, can also have pronounced influences on the viticulture of an area.

Mediterranean Climates

Wine regions with Mediterranean climates are characterized by their long growing seasons of moderate to warm temperatures. Throughout the year there is little seasonal change with temperatures in the winter generally warmer than those of maritime and continental climates. During the grapevine growing season, there is very little rain fall (with most precipitation occurring in the winter months) which increases the risk of the viticultural hazard of drought and may present the need for supplemental irrigation.

The Mediterranean climate is most readily associated with the areas around the Mediterranean basin, where viticulture and winemaking first flourished on a large scale due to the influence of the Phoenicians, Greeks and Romans of the ancient world.

Wine Regions with Mediterranean Climates

- Tuscany and most other Italian wine regions
- Most Greek wine regions
- Most Israeli wine regions
- Most Lebanese wine regions
- Southern Rhone Valley
- Catalonia
- Languedoc
- Provence
- Coastal Portuguese wine region
- Primorska Slovenian wine region
- Napa Valley and other coastal California wine regions
- Western Australia and South Australia wine regions
- Chilean Central Valley
- Coastal South African wine regions
- Western coastal Turkish wine regions

Continental Climates

Wine regions with continental climates are characterized by the very marked seasonal changes that occur throughout the growing season, with very hot temperatures during the summer season and very cold winters. This is generally described as having a high degree of continentality. Regions with this type of climate are often found inland on continents without a significant body of water, such as aninland sea, that can moderate their temperatures. Often during the growing season, continental climates will have wide diurnal temperature variations with very warm temperatures during the day that drop drastically at night. During the winter and early spring months, frost and hail can be viticultural hazards. Depending on the particular macroclimate of the region, irrigation may be needed to supplement seasonal rainfall. These many climatic influences contribute to the wide vintage variation that is often typical of continental climates such as Burgundy. There are more wine regions with continental climates in the northern hemisphere than there are in the southern hemisphere. This is due, in part, to small land mass size of southern hemisphere continents relative to the large oceans nearby. This difference means that the oceans exert a more direct influence on the climate of the southern hemisphere wine regions (making them maritime or possibly Mediterranean) than they would on the larger northern hemisphere continents.

There are also several wine regions (such as Spain) that have areas that exhibit a continental Mediterranean climate due to their altitude or distance from the sea. These regions will have more distinct seasonal change than Mediterranean climates, but still retain some characteristics like a long growing season that is very dry during the summer.

Wine Regions with Continental Climates

- Burgundy
- Rioja
- Piedmont
- Cote-Rotie and other Northern Rhone wine regions
- Douro
- Most of the Loire Valley
- Most Austrian wine regions
- Most Czech wine regions
- Most Hungarian wine regions
- Most Romanian wine regions
- Most Russian wine regions
- Most Turkish wine regions

- Columbia Valley
- Most Canadian wine regions
- Mendoza

Maritime Climates

Wine regions with maritime climates are characterized by their close proximity to large bodies of water (such as oceans, estuaries and inland seas) that moderate their temperatures. Maritime climates share many characteristics with both Mediterranean and continental climates and are often described as a "middle ground" between the two extremes. Like Mediterranean climates, maritime climates have a long growing season, with water currents moderating the region's temperatures. However, while Mediterranean climates are usually very dry during the growing season, maritime climates are often subject to the viticultural hazards of excessive rain and humidity which may promote various grape diseases, such as mold and mildew. Like continental climates, maritime climates will have distinct seasonal changes, but they are usually not as drastic, with warm, rather than hot, summers and cool, rather than cold, winters.

Wine Regions with Maritime Climates

- Bordeaux
- Muscadet
- Rias Baixes
- Willamette Valley
- Long Island
- Most New Zealand wine regions
- Southern Chile
- Itata Valley
- Biobío Valley
- Malleco Valley

Aridity Index

An aridity index (AI) is a numerical indicator of the degree of dryness of the climate at a given location. A number of aridity indices have been proposed; these indicators serve to identify, locate or delimit regions that suffer from a deficit of available water, a condition that can severely affect the effective use of the land for such activities as agriculture or stock-farming.

Historical Background and Indices

At the turn of the 20th century, Wladimir Köppen and Rudolf Geiger developed the concept of a climate classification where arid regions were defined as those places where the annual rainfall accumulation (in centimetres) is less than $R/2$, where:

- $R = 2 \times T$ if rainfall occurs mainly in the cold season,
- $R = 2 \times T + 14$ if rainfall is evenly distributed throughout the year, and
- $R = 2 \times T + 28$ if rainfall occurs mainly in the hot season.

where T is the mean annual temperature in Celsius.

This was one of the first attempts at defining an aridity index, one that reflects the effects of the thermal regime and the amount and distribution of precipitation in determining the native vegetation possible in an area. It recognizes the significance of temperature in allowing colder places such as northern Canada to be seen as humid with the same level of precipitation as some tropical deserts because of lower levels of potential evapotranspiration in colder places. In the subtropics, the allowance for the distribution of rainfall between warm and cold seasons recognizes that winter rainfall is more effective for plant growth that can flourish in the winter and go dormant in the summer than the same amount of summer rainfall during a warm-to-hot season. Thus a place like Athens, Greece that gets most of its rainfall in winter can be considered to have a humid climate (as attested in lush foliage) with roughly the same amount of rainfall that imposes semi-desert conditions in Midland, Texas, where rainfall largely occurs in the summer.

In 1948, C. W. Thornthwaite proposed an AI defined as:

$$AI_T = 100 \times \frac{d}{n}$$

where the water deficiency d is calculated as the sum of the monthly differences between precipitation and potential evapotranspiration for those months when the normal precipitation is less than the normal evapotranspiration; and where n stands for the sum of monthly values of potential evapotranspiration for the deficient months (after Huschke, 1959). This AI was later used by Meigs (1961) to delineate the arid zones of the world in the context of the UNESCO Arid Zone Research programme.

In the preparations leading to the UN Conference on Desertification (UNCOD), the United Nations Environment Programme (UNEP) issued a dryness map based on a different aridity index, proposed originally by Mikhail Ivanovich Budyko (1958) and defined as follows:

$$AI_B = 100 \times \frac{R}{LP}$$

where R is the mean annual net radiation (also known as the net radiation balance), P is the mean annual precipitation, and L is the latent heat of vaporization for water. Note that this index is dimensionless and that the variables , and can be expressed in any system of units that is self-consistent. More recently, the UNEP has adopted yet another index of aridity, defined as:

$$AI_U = \frac{P}{PET}$$

where PET is the potential evapotranspiration and P is the average annual precipitation (UNEP, 1992). Here also, PET and P must be expressed in the same units, e.g., in milimetres. In this latter case, the boundaries that define various degrees of aridity and the approximate areas involved are as follows:

Classification	*Aridity Index*	*Global land area*
Hyperarid	AI < 0.05	7.5%
Arid	0.05 < AI < 0.20	12.1%
Semi-arid	0.20 < AI < 0.50	17.7%
Dry subhumid	0.50 < AI < 0.65	9.9%

Holdridge Life Zones

The Holdridge life zones system is a global bioclimatic scheme for the classification of land areas. It was first published by Leslie Holdridge in 1947, and updated in 1967. It is a relatively simple system based on few empirical data, giving objective mapping criteria. A basic assumption of the system is that both soil and climax vegetation can be mapped once climate is known.

While it was first designed for tropical and subtropical area, the system applies globally. The system has been shown to fit tropical vegetation zones, Mediterranean zones, and borealzones, but is less applicable to cold oceanic or cold arid climates where moisture becomes the determining factor. The system has found a major use in assessing the possible changes in natural vegetation patterns due to global warming.

The three axes of the barycentric subdivisions are:

- precipitation (annual, logarithmic)
- biotemperature (mean annual, logarithmic)
- potential evapotranspiration ratio (PET) to mean total annual precipitation.

Further indicators incorporated into the system are:

- humidity provinces
- latitudinal regions
- altitudinal belts

Biotemperature is based on the growing season length and temperature. It is measured as the mean of all temperatures above freezing, with all temperatures below freezing adjusted to 0 °C, as plants are dormant at these temperatures. Holdridge's system uses biotemperature first, rather than the temperate latitude bias of Merriam's life zones, and does not primarily consider elevation. The system is considered more appropriate to the complexities of tropical vegetation than Merriam's system.

Classes

The classes defined within the system, as used by IIASA, are:

1. Polar desert
2. Subpolar dry tundra
3. Subpolar moist tundra
4. Subpolar wet tundra
5. Subpolar rain tundra
6. Boreal desert
7. Boreal dry scrub
8. Boreal moist forest
9. Boreal wet forest
10. Boreal rain forest
11. Cool temperate desert
12. Cool temperate desert scrub
13. Cool temperate steppe
14. Cool temperate moist forest
15. Cool temperate wet forest
16. Cool temperate rain forest
17. Warm temperate desert

18. Warm temperate desert scrub
19. Warm temperate thorn scrub
20. Warm temperate dry forest
21. Warm temperate moist forest
22. Warm temperate wet forest
23. Warm temperate rain forest
24. Subtropical desert
25. Subtropical desert scrub
26. Subtropical thorn woodland
27. Subtropical dry forest
28. Subtropical moist forest
29. Subtropical wet forest
30. Subtropical rain forest
31. Tropical desert
32. Tropical desert scrub
33. Tropical thorn woodland
34. Tropical very dry forest
35. Tropical dry forest
36. Tropical moist forest
37. Tropical wet forest
38. Tropical rain forest

4

Humid Continental Climate

A humid continental climate (Köppen prefix *D* and a third letter of *a* or *b*) is a climatic region typified by large seasonal temperature differences, with warm to hot (and often humid) summers and cold (sometimes severely cold) winters.

Precipitation is relatively well-distributed year-round in many areas with this climate, while others may see a marked reduction in wintry precipitation and even a wintertime drought. Snowfall, regardless of average seasonal totals, occurs in all areas with a humid continental climate and in many such places is more common than rainduring the height of winter. In places with sufficient wintertime precipitation, the snow cover is often deep. Most summer rainfall occurs during thunderstorms and a very occasional tropical system. Though humidity levels are often high in locations with humid continental climates, it is important to note that the "humid" designation does not mean that the humidity levels are necessarily high, but that the climate is not dry enough to be classified assemi-arid or arid. Very few areas with a humid continental climate fall in the *Dsa* or *Dsb* categories; generally these are adjacent to Mediterranean climates where the elevation precludes such classification due to colder winters.

Humid continental climates tend to be found above 40° N latitude, within the central and northeastern portions ofNorth America, Europe, and Asia. They are much less commonly found in the Southern Hemisphere due to the larger ocean area at that latitude and the consequent greater maritime moderation. The Köppen definition of this climate regarding temperature is as follows: the mean temperature of the coldest month must be below −3 °C (26.6 °F) (some climatologists prefer to use the freezing mark), and there must be at least four months whose mean temperatures are at or above 10 °C (50 °F). In addition, the location in question must not be semi-arid or arid.

Explanation of Lettering

Under Köppen, the following variants of this climate are possible:

- The second letter
- *w*: a dry winter — the driest winter month has at most one-tenth of the precipitation found in the wettest summer month.
- *s*: a dry summer — the driest summer month has at most 30 millimetres (1.18 in) of rainfall and has at most S! the precipitation of the wettest winter month.
- *f*: Does not meet either of the above specifications.
- The third letter
- *a*: Warmest month averages above 22 °C (71.6 °F)
- *b*: Does not meet the requirements for *a*, but there still are at least four months above 10 °C (50 °F).

Dfa/Dwa/Dsa: Hot (or very warm) Summer Subtype

A hot (or very warm) version of a continental climate features an average temperature of at least 22 °C (71.6 °F) in its warmest month. The warmest month is usually July, though in some cases, it can be August. In this region, July afternoon temperatures average up to 32 °C (90 °F), while the January mean temperature can be far below – 3 °C (26.6 °F). In Europe, it is found in areas largely wrapping the Black Sea coastline from Romania, Ukraine, along the Danube River (parts of Hungary and Serbia) and in the Don River estuary in Russia. In Asia, around the Caspian Sea in Russia, Kazakhstan, in parts of Iran, Turkey and parts of Kashmir in India. It covers a great swath of Northeast China, almost all of North Korea, most of South Korea and much of northern Japan as well as various central areas.

In East Asia, this climate exhibits a monsoonal tendency with much higher precipitation in summer than in winter, and due the effects of the strong Siberian High much colder winter temperatures than similar latitudes around the world, however with lower snowfall, the exception being western Japan with its heavy snowfall. Within North America it includes parts of southern New England and the Middle Atlanticstates, much of the Midwestern United States and small parts of Southern Ontario, Canada from the Atlantic to the 100th meridian and in the general range of between 39 °N to 44 °N latitude (with a larger north-south spread in the western portion due to the lack of maritime influences); precipitation increases further eastward in this zone and is less seasonally uniform in the west.

The 0 °C (32 °F) isotherm (freeze line) and the –3 °C (27 °F) isotherm (persistent-snow line) are often both argued as the statistical dividing line between the humid continental climate dominating areas to the north and west and the humid subtropical climate dominating areas to the south and southwest. The Köppen climate classification, the most popular climate classification, uses –3 °C (27 °F) as its lower threshold criterion; however, many climatologists in the U.S. prefer to use 0 °C (32 °F) as the standard because they feel it better reflects consistency in regional floristic character (i.e. forest composition/type). Between these lines lies a gradual zone of transition between the two climates.

The western states of the central United States (namely Montana, Wyoming, parts of southern Idaho, parts of Colorado, western Nebraska, and parts of western North and South Dakota) have thermal regimes which fit the *Dfa* climate type, but are quite dry, and are generally grouped with the steppe (*BSk*) climates.

Outside of North America the *Dfa* climate type is present near the Black Sea in southern Ukraine, the Southern Federal District of Russia, southern Moldova, and parts of southern and western Romania, but tends to be drier and can be even semi-arid, in these places. Tohoku in Japanbetween Tokyo and Hokkaidô also has a climate with Köppen classification *Dfa*, but is wetter even than that part of North America with this climate type. A variant which has dry winters and hence much lower snowfall with monsoonal type summer rainfall is to be found in north-eastern China including coastal regions of the Yellow Sea and over much of the Korean Peninsula; it has the Köppen classification *Dwa*. Much of central Asia, northwestern China, and southern Mongolia have a thermal regime similar to that of the *Dfa* climate type, but these regions receive so little precipitation that they are more often classified as steppes (*BSk*) or deserts (*BWk*).

It appears nowhere within the Southern Hemisphere, which has no large landmasses so situated in the middle latitudes that allow the combination of hot summers and at least one month of sub-freezing temperatures.

Dfb/Dwb/Dsb: Warm Summer Subtype

The warm summer version of the humid continental climate covers a much larger area than the hot subtype. In North America, the climate zone covers from about 44°N to 50°N latitude mostly east of the 100th meridian. However, it can be found as far north as 54°N, and further west in the Canadian Prairie Provinces and below 40°N in the high Appalachians. In Europe this subtype reaches its most northerly latitude at nearly 61° N. Areas featuring this subtype of the continental climate have an average temperature in the warmest month below 22°C. Summer high temperatures in this zone typically average between 21–28 °C (70–82 °F) during the daytime and the average temperatures in the coldest month are generally far below the–3 °C (27 °F) isotherm.

Such high-altitude locations as South Lake Tahoe, California and Aspen, Colorado in the western United States exhibit local Dfb climates. The south-central and southwestern Prairie Provinces also fits the *Dfb* criteria from a thermal profile, but because of semi-arid precipitation portions of it are grouped into the *BSk* category.

In Europe, it is also found in central Scandinavia. In eastern Central Europe (eastern Austria, eastern Germany, Poland, Czech Republic, Slovakia, Hungary, Ukraine, northern Romania) and in coastal areas of central Scandinavia is a warm summer subtype with less severe winters, more similar to the winters of the hot summer subtype found in eastern North America – the winters here are modified by the oceanic climate influence of western Europe.

The warm summer subtype is marked by mild summers, long cold winters and less precipitation than the hot summer subtype, however, short periods of extreme heat are not uncommon. Northern Japan has a similar climate. Much of Mongolia and parts of southern Siberia have a thermal regime fitting this climate, but they have steppe- or desert-like precipitation, and so are not really considered to have a humid continental climate. In the Southern Hemisphere it exists in well-defined areas only in the Southern Alps of New Zealand and perhaps as isolated microclimates of the southern Andes of Chile and Argentina.

Humid Subtropical Climate

A humid subtropical climate (Köppen climate classification *Cfa* or *Cwa*) is a climate zone characterized by hot, humid summers and generally mild to cool winters. Under the Köppen climate definition, this category of climate type covers a broad range of attributes, and the term "subtropical" may be a misnomer for locations along the outer ranges.

Description

The Köppen definition of this climate is for the coldest month's mean temperature to be between "3 °C (26.6 °F), although some climatologists prefer to use 0 °C (32 °F), and 18 °C (64.4 °F), and the warmest month to be above 22 °C (71.6 °F). It is either accompanied with a dry winter (Köppen: w) — or has no distinguished dry season (Köppen: f).

Significant amounts of precipitation occur in all seasons in most areas, and though in regions bordering onsemi-arid climates (usually at the western margins), irregular droughts can be common and catastrophic toagriculture. Winter rainfall (and sometimes snowfall) is associated with large storms that the westerlies steer from west to east. Most summer rainfall occurs during thunderstorms and an occasional tropical storm, hurricane or cyclone.

Overview Distribution

Humid subtropical climates normally lies on the southeast side of all continents, generally between latitudes 25° and 40° north and tend towards coastal or near costal strips but in some cases extend well inland, most notably in China and the United States. Using the Koppen climate zones, in North America it covers a wide swath of the southeastern part of the United States extending inland to the middle Mississipi River Valley, including a large section of the mid-Atlantic coast up to as far north as the Philadelphia and New York City metropolitan areas. Most of Florida is subtropical but the far south has either a Tropical savanna climate (wet/dry) or Tropical monsoon climate climate. In Mexico, the subtropical zone covers small areas scattered around the northeastern part of the country, in close proximity to the Gulf of Mexico.

In South America, this climate extends over a few states of southern Brazil, including Paraná (state), into sections of Paraguay, all of Uruguay, and the Río de la Plata region in Argentina.

In Europe, primarly due to the milder winters it extends above latitude 45° N, covering regions such as the Po Valley of Italy, which includes Venice, the Toulouse region of France, and in places along the Adriatic and Black Sea coasts which are too wet for inclusion in the Mediterrean climate schema, inland from these areas there are isolated pockets where the climate is borderline subtropical but these zones are usually classed as oceanic or humid continental. Average summer temperatures in areas of Europe with this climate are generally not as hot as most other subtropical zones around the world, but the growing season can be adequately long.

In Africa, over narrow coastal sections of southern and eastern South Africa, primarly in KwaZulu-Natal province and the adjacent border area of Mozambique.

In Southwestern Asian, in the Gilan of Iran, in parts of the Caucasus, in Azerbaijan and Georgia wedged between the Caspian Sea and Black seas, in small areas of the southern Russian Federation and coastal (Black Sea) Turkey. In India, it covers a small area of the Punjab region but with untypical, more extreme conditions than those found in most subtropical climates worldwide and has characteristics of a monsoon climate, without the tropical conditions during winter.

In East Asia, the influence of the strong Siberian anticyclone brings colder winter temperatures southward, pushing the southern boundary of this regime to between the Yangtze and Yellow River valleys in China. Its range is further north over the Pacific coastal regions of the Japanese archipelago. At Hainan Island, the southernmost island in China at latitude 20° Nand in Taiwan, the climate transitions from subtropical into fully tropical.

In Australia, in parts of both coastal and inland southern Queensland, New South Wales and a much smaller area in Victoria state fall into this climate grouping.

Africa

In Africa, the humid subtropical climates are found in two separate areas on the southern hemisphere of the continent. The *Cwa* climate is found in over a large portion of the interior of the Middle and Eastern African regions. This area includes; central Angola, northeastern Zimbabwe, the Niassa, Manica and Tete provinces of Mozambique, the southern Congo provinces, southwest Tanzania, and the majority of Malawi, and Zambia. Some lower portions of the Ethiopian Highlands also have this climate.

The *Cfa* climate covers a relatively small area of coastal KwaZulu-Natal and Eastern Cape provinces of South Africa, and is characterised by oceanic influences that give mild temperatures, especially in winter when temperatures do not drop as low as in many other regions within the humid subtropical category.

Asia

Locations in Asia with a humid subtropical climate differ from those in other continents in that they often have marked seasonal differences in precipitation, if not very dry winters.

East Asia

In East Asia, this type climate is found in northern Vietnam, the southeastern quarter of mainland China, the northern half of Taiwan, narrow areas along the coast of South Korea, and Japan (Kyushu, Shikoku, and most of Honshu). Cities on the equatorward boundary of this zone include Hong Kong and Taipei while Qingdao is on the northern boundary.

In most of this region, there is extremely limited precipitation during the winter, owing to the powerful anticyclonic winds from Siberia. Only in inland areas below the Yangtze River and coastal areas between approximately the Huai River and the beginning of the coast of Guangdong is there sufficient winter rainfall to produce a *Cfa* climate; even in these areas, rainfall and streamflow show a highly pronounced summer peak quite unlike other regions of this climate type.

The only area where winter precipitation equals the summer rain is on the "San-in" (Sea of Japan), or western, coast of Japan, which during winter is on the windward side of the westerlies. The winter precipitation in these regions is usually produced by low-pressure systems off the east coast that develop in the onshore flow from the Siberian high. Summer rainfall comes from the East Asian Monsoon and from frequent typhoons. Annual rainfall is generally over 1,000 mm (40 inches), and in areas below the Himalayas can be much higher still.

South Asia

Humid subtropical climates can also be found in South Asia, primarily in northern India. However, the humid subtropical climates exhibited here differ markedly from humid subtropical climates in East Asia (and for that matter a good portion of the globe). Winters here are typically mild, dry and relatively short. They also tend to be foggy. Summers tend to be long and very hot, with high temperatures sometimes exceeding 40°C.

They also tend to be extremely dry, complete with dust storms, traits usually associated with arid or semiarid climates. This is followed by themonsoons, where the region experiences heavy rain on almost a daily basis. Average high temperatures decreases during the monsoon season but the humidity increases. This results in hot and humid conditions, similar to summers.in humid subtropical climates. Cities such as New Delhi, Lucknow and Kanpur exhibit this atypical version of the climate. Islamabad also features this weather pattern, but with wetter winters.

In South Asia, humid subtropical climates generally border on continental climates as altitude increases, or on winter-rainfall climates in Pakistan.

Southwestern Asia (Northern Middle East and Caucasus)

Although humid subtropical climates in Asia are mostly confined to the southeastern quarter of the continent, there are areas on the Caspian Sea and Black Sea with humid subtropical climates that are unusually warm for their high latitudes and also unusual for this climate type, that snowfall in winter is relatively common, but is usually of a short duration.

In the narrow Caspian coastal strip of Iran (Gilan and Mazandaran) a humid subtropical climate prevails. Annual rainfall ranges from around 740 mm (29 inches) at Sari to over 2,000 mm (78 inches) at Bandar-e Anzali, and is heavy throughout the year, with a maximum in October or November when Bandar-e Anzali can average 400 millimetres (16 inches). Temperatures are generally moderate in comparison with other parts of Southwestern Asia. In Rasht, the average maximum in July is around 28 °C (82 °F) but with near-saturation humidity, whilst in January it is around 9 °C (48 °F). The heavy, evenly distributed rainfall extends north into the Caspian coastal strip of Azerbaijan up to its northern border but this climate in Azerbaijan is, however, a *Cfb*/*Cfa* (*Oceanic climate*/*Humid subtropical climate*) borderline case. During winter, the coastal areas can receive snowfall, but is usually of a short duration. Annual rainfall in Lankaran in the southeast averages up to 1,800 mm (70 inches) and is heavy throughout the year; and annual rainfall is generally over 1,000 mm (40 inches) in the foothills of the Caucasus in the northeast, as altitude increases and the humid subtropical climate changes to the oceanic climate

Western Georgia in the Kolkheti Lowland and the north coast of Turkey, have a climate similar to that of Gilan and Mazandaran in Iran and very similar to that of southeastern and northern Azerbaijan. Temperatures range from 22 °C in summer to 5 °C in winter and rainfall is even heavier than in Caspian Iran, up to 2,300 millimetres per year in Hopa (Turkey) and up to 2,718 millimetres per year in Batumi (Georgia) falling throughout the year. This climate in northern Turkey and western Georgia is, again, a *Cfb*/*Cfa* (*Oceanic climate*/*Humid subtropical climate*) borderline case. And again, during winter, the coastal areas can receive snowfall, but is usually of a short duration.

North America

In North America, humid subtropical climates are almost exclusively the domain of the American South, including the following states: the eastern half of Texas, Oklahoma, Louisiana, Arkansas, Alabama, Mississippi, North Carolina, South Carolina, Tennessee, Georgia,

Kentucky, most of Florida, Virginia, and lower elevation areas of the southwest and Eastern Panhandle regions of West Virginia.

The climate in many of these states is subject to extremes. The humid subtropical climate can also be found in the Mid-Atlantic, primarily Maryland, Delaware, the District of Columbia, southeastern Pennsylvania, southern New Jersey and far southern New York, specifically New York City and sections of Long Island. It can also be found in the Midwest, primarily in the central and southern portions of Kansas and Missouri, and the southern portions of Illinois, Indiana and Ohio. The Mid-Atlantic and Midwestern areas included in this climate typically see snowfall during the winter, with occasional heavy storms. The archetypal humid subtropical climate is best exemplified by the American Deep South, because the summers are long and almost tropical, and temperatures reach freezing only a few times in the winter with rare snowfall, usually three inches or less. Summers in this zone are hot and humid, with daily averages above 25 °C (77 °F) with average daily maximums above 30 °C (86 °F) .

In Mexico, there are small areas of *Cfa* and *Cwa* climates. They are both caused by the high elevations of Trans-Mexican Volcanic Belt and Sierra Madre Oriental. Despite being located at higher elevations, these locations have summers that are too warm to qualify as a subtropical highland climate. Guadalajara's climate is a major example of this.

Characteristics and Variants

Outside of the isolated higher-altitude sections of Mexico, the southernmost limits of this climate in North America lie just north of South Floridaand around southern coastal Texas. Cities at the southernmost limits of this climate, such as Orlando and ampa generally feature warm weather year round and minimal temperature differences between seasons. These cities fall just short of having a true tropical climate. In contrast, cities at the northernmost limits of the humid subtropical region, such as New York City, Philadelphia, and Louisville, Kentucky feature winters that are barely warm enough to qualify as a humid subtropical climate. These cities generally experience much greater seasonal variation, featuring hot, humid summers and cold winters. Areas farther north than this, inland, or at a higher elevation, fall into the humid continental climate category with harsher winters.

Snowfall varies greatly in this climate zone. In locations at the southern limits of this zone and areas around the Gulf Coast, cities such as Orlando, Tampa, Houston, and New Orleans rarely see snowfall, which occurs, at most, a few times per generation. In inland southern cities farther north, such as Birmingham, Atlanta, Memphis, Little Rock, Nashville, Dallas, Charlotte, and Raleigh, snow typically falls once or twice a season and is usually three inches or less. Ice storms are not unusual at these locations. However for the majority of the winter here, temperatures remain above or well above freezing, with slight plant growth. In the northern limits of this climate zone, cities such as Philadelphiaand New York City experience snow every winter, sometimes accumulating heavily although it melts more quickly than in regions to the north.

Precipitation is plentiful in the humid subtropical climate zone in North America. Although most areas tend to have precipitation spread evenly throughout the year, a somewhat monsoon-like pattern is seen in parts of the Southeast (in locales such as Augusta, Georgia and Columbia, South Carolina), which experience dry winters (by humid subtropical standards) and warm springs, followed immediately by a long, hot, rainy and humid summer. In addition, areas in Texas that are slightly inland from the Gulf of Mexico, such as Austin and San Antonio that border the semi-arid climate zone, generally see a peak of precipitation in the spring, and a deep, drought-like nadir in mid-summer.

South America

Humid subtropical climates are found in a sizeable portion of South America. Most of north-eastern Argentina, Uruguay, southern Brazil, and eastern Paraguay features this climate. Major cities such as São Paulo, Buenos Aires, Porto Alegre and Montevideo have a humid subtropical climate, generally in the form of hot humid summers and mild to cool winters. These areas, which include the Pampas, generally feature a *Cfa* climate categorization. The *Cwa* climate occurs in parts of tropical highlands of São Paulo state, Minas Gerais and near the Andean highland in northwestern Argentina. These highland areas feature summer temperatures that are warm enough to fall outside the subtropical highland climate category.

Australia

This zone contains the only regions where soils are not acutely deficient in phosphorus, as well as the heaviest rainfall south of the Tropic of Capricorn, making it prime agricultural country, centred on towns such as Coffs Harbour, Grafton, Kempsey, Port Macquarie, Tamworth, and Moree.

Variations in Australia

There is variation in climate within this zone. Annual rainfall on the coast can reach as high as 2,000 mm (80 inches) in favourable locations and is generally above 1,000 mm (40 inches). However, because most of the heaviest two- and three-day rainfalls in the world occur in this coastal zone as a result of east coast lows forming to the north of a large high pressure system, there can be great variation in rainfall from year to year. At Lismore in the centre of this zone, the annual rainfall can range from less than 550 mm (22 inches) in 1915 to more than 2,780 mm (110 inches) in 1950. There is usually a distinct summer rainfall maximum that becomes more pronounced moving northwards: in Brisbane the wettest month (February) receives five times the rainfall of the driest (September). Temperatures are very warm to hot but not excessive: the average maximum in February is usually around 29 °C (84 °F) and in July around 21 °C (70 °F). Frosts are extremely rare except at higher elevations, but temperatures over 35ÚC (95ÚF) are not common on the coast.

North of the *Cfa* climate zone there is a zone centred upon Rockhampton and extending up to the Atherton Tableland of Köppen Cwaclimate. This has a very pronounced dry winter with often negligible rainfall between June and October, and winter temperatures generally only slightly below 18°C, above which one would have a tropical savanna, or Aw, climate.

Europe

In other climate classification systems outside of the Köppen, most of the following places would not be included the humid subtropical grouping. Also higher precipitation and high humidity of summer is not present nearly to the degree that it is in subtropical regions of North America and Asia, making its distinction in Europe all the more difficult.

Some areas of Europe, such as parts of the northeastern interior of the Iberian Peninsula, southern France Garonne Valley, Po River Valley and Adriatic northern Italy, parts of Epirus in Greece around the area of Ioannina, parts of coastal northern Croatia, coastal Slovenia fall into this classification. Along the Black Sea coast of Bulgaria, Romania, Sochi, Russia and southernmost Ukraine have summers too warm (>22°C in the warmest month) to qualify as oceanic, no freezing month, and enough summer precipitation and sometimes humid conditions to preclude their classification as Mediterranean but rather border on or are sometimes defined as Humid continental climates. All these areas are subject to occasional, in some cases repeated snowfalls and freezes during winter. In the Azores, some islands have this climate, with very mild and rainy winters (> 13°C) and no snowfall, hot summers (> 22 or 23°C) but with no dry season during the warmest period, which means that they can be classified neither as oceanic, nor as Mediterranean, but only as humid subtropical climate, as with Corvo Island.

Humidity

Humidity is a term for water vapor in the air, and can refer to any one of several measurements of humidity. Formally, humid air is not "moist air" but a mixture of water vapor and other constituents of air, and humidity is defined in terms of the water content of this mixture, called the *Absolute humidity*. In everyday usage, it commonly refers to *relative humidity*, expressed as a percent in weather forecasts and on household humidistats; it is so called because it measures the current absolute humidity relative to the maximum. *Specific humidity* is a ratioof the water vapor content of the mixture to the total air content (on a mass basis). The water vapor content of the mixture can be measured either as mass per volume or as a partial pressure, depending on the usage.

In meteorology, relative humidity indicates the likelihood of precipitation, dew, or fog. High relative humidity reduces the effectiveness of sweating in cooling the body by reducing the rate of evaporation of moisture from the skin. This effect is calculated in a heat index table, used during summer weather.

Absolute Humidity

Absolute humidity is an amount of water vapor, usually discussed per unit volume. The mass of water vapor, m_w, per unit volume of total moist air, V_{net}, can be expressed as follows:

$$AH = \frac{m_w}{V_{net}}.$$

Absolute humidity in air ranges from zero to roughly 30 grams per cubic meter when the air is saturated at 30 °C.

The absolute humidity changes as air pressure changes. This is very inconvenient for chemical engineering calculations, e.g. for clothes dryers, where temperature can vary considerably. As a result, absolute humidity is generally defined in chemical engineering as mass of water vapor per unit mass of dry air, also known as the mass mixing ratio, which is much more rigorous for heat and mass balance calculations. Mass of water per unit volume as in the equation above would then be defined as volumetric humidity. Because of the potential confusion, British Standard BS 1339 (revised 2002) suggests avoiding the term "absolute humidity". Units should always be carefully checked. Most humidity charts are given in g/kg or kg/kg, but any mass units may be used. The field concerned with the study of physical and thermodynamic properties of gas-vapor mixtures is named Psychrometrics.

Relative Humidity

Relative humidity is a term used to describe the amount of water vapor in a mixture of air and water vapor. It is defined as the ratio of the partial pressure of water vapor in the air-water mixture to the saturated vapor pressure of water at those conditions. The relative humidity of air depends not only on temperature but also on pressure of the system of interest.

Relative humidity is normally expressed as a percentage and is calculated by using the following equation, it is defined as the ratio of the partial pressure of water vapor (H2O) (ew) in the mixture to the saturated vapor pressure of water $\left(e^{*}_{w}\right)$ at a prescribed temperature.

$$\phi = \frac{e_w}{e^{*}_{w}} \times 100\%$$

Relative humidity is an important metric used in weather forecasts and reports, as it is an indicator of the likelihood of precipitation, dew, or fog. In hot summer weather, a rise in relative humidity also increases the apparent temperature to humans (and other animals) by hindering the evaporation of perspiration from the skin as the relative humidity rises. For example, according to the Heat Index, a relative humidity of 75% at 80°F (27°C) would feel like 83.574°F ±1.3 °F (28.652°C ±0.7 °C) at ~44% relative humidity.

Specific Humidity

Specific humidity is the ratio of water vapor to dry air in a particular mass, and is sometimes referred to as humidity ratio. Specific humidity ratio is expressed as a ratio of mass of water vapor, m_u, per unit mass of dry air m_a.

That ratio is defined as:

$$SH = \frac{m_v}{m_a}.$$

Specific humidity can be expressed in other ways including:

$$SH = \frac{0.622 p_{(H_2O)}}{p_{(dryair)}}$$

or:

$$SH = \frac{0.622 p_{(H_2O)}}{p - p_{(H_2O)}}.$$

Using the definition of specific humidity, the relative humidity can be expressed as

$$\phi = \frac{SH * p}{(0.622 + SH) p^*_{(H_2O)}} \times 100$$

However, specific humidity is also defined as the ratio of water vapor to the total mass of the system in meteorology. "Mixing ratio" is used to name the definition in this section beginning.

Measurement

There are various devices used to measure and regulate relative humidity. A device used to measure relative humidity is called a psychrometer or hygrometer. A humidistat is used to regulate the relative humidity of a building with a dehumidifier. These can be analogous to thermometer and thermostat for temperature control.

Humidity is also measured on a global scale using remotely placed satellites. These satellites are able to detect the concentration of water in the troposphere at altitudes between 4 and 12 kilometres. Satellites that can measure water vapor have sensors that are sensitive to infrared radiation. Water vapor specifically absorbs and re-radiates radiation in this spectral band. Satellite water vapor imagery plays an important role in monitoring climate conditions (like the formation of thunderstorms) and in the development of future weather forecasts.

Climate

While humidity itself is a climate variable, it also interacts strongly with other climate variables. The humidity is affected by winds and by rainfall. At the same time, humidity affects the energy budget and thereby influences temperatures in two major ways. First, water vapor in the atmosphere contains "latent" energy. During transpiration or evaporation, this latent heat is removed from surface liquid, cooling the earth's surface. This is the biggest non-radiative cooling effect at the surface. It compensates for roughly 70% of the average net radiative warming at the surface. Second, water vapor is the most important of all greenhouse gases. Water vapor, like a green lens that allows green light to pass through it but absorbs red light, is a "selective absorber". Along with other greenhouse gases, water vapor is transparent to most solar energy, as you can literally. But it absorbs the infrared energy emitted (radiated) upward by the earth's surface, which is the reason that humid areas experience very little nocturnal cooling but dry desert regions cool considerably at night. This selective absorption causes the greenhouse effect. It raises the surface temperature substantially above its theoretical radiative equilibrium temperature with the sun, and water vapor is the cause of more of this warming than any other greenhouse gas.

The most humid cities on earth are generally located closer to the equator, near coastal regions. Cities in South and Southeast Asia are among the most humid, such as Kolkata, Chennai and Cochin in India, the cities of Manila in the Philippines and Bangkok in Thailand: these places experience extreme humidity during their rainy seasons combined with warmth giving the feel of a lukewarm sauna. Darwin, Australia experiences an extremely humid wet season from December to April. Shanghai and Hong Kong in China also have an extreme humid period in their summer months. Kuala Lumpur and Singapore have very high humidity all year round because of their proximity to water bodies and the equator and overcast weather. Perfectly clear days are dependent largely upon the season in which one decides to travel. During the South-west and North-east Monsoon seasons (respectively, late May to September and November to March), expect heavy rains and a relatively high humidity post-rainfall. Outside the monsoon seasons, humidity is high (in comparison to countries North of the Equator), but completely sunny days abound. In cooler places such as Northern Tasmania, Australia, high humidity is experienced all year due to the ocean between mainland Australia and Tasmania. In the summer the hot dry air is absorbed by this ocean and the temperature rarely climbs above 35 degrees Celsius.

In the United States the most humid cities, strictly in terms of relative humidity, are Forks and Olympia, Washington. This fact may come as a surprise to many, as the climate in this region rarely exhibits the discomfort usually associated with high humidity. Dew points are typically much lower on the West Coast than on the East. Because high dew points play a more significant role than relative humidity in the discomfort created during humid days, the air in these western cities usually does not feel "humid". The highest dew points in the US are found in coastal Florida and Texas. When comparing Key West and Houston, two of the most humid cities from those states, coastal Florida seems to have the higher dew points on average. However, Houston lacks the coastal breeze present in Key West, and, as a much larger city, it suffers from the urban heat islandeffect. A dew point of 86 degrees Fahrenheit was recorded in southern Minnesota on July 23, 2005, though dew points over 80 degrees Fahrenheit are rare there. The US city with the lowest annual relative humidity is Las Vegas, Nevada, averaging 39% for a high and 21% as a low.

Air Density and Volume

Relative humidity depends on water vaporization and condensation, which, in turn, mainly depends on temperature. Therefore, when applying more pressure to a gas saturated with water, all components will initially decrease in volume approximately according to the ideal gas law. However, some of the water will condense until returning to almost the same relative humidity as before, giving the resulting total volume deviating from what the ideal gas law predicted. Conversely, decreasing temperature would also make some water condense, again making the final volume deviating from predicted by the ideal gas law. Therefore, gas volume may alternatively be expressed as the dry volume, excluding the humidity content. This fraction more accurately follows the ideal gas law. On the contrary the saturated volume is the volume a gas mixture would have if humidity was added to it until saturation (or 100% relative humidity).

Humid air is less dense than dry air because a molecule of water (M H– 18 u) is less massive than either a molecule of nitrogen (M H– 28) or a molecule of oxygen (M H– 32). About 78% of the molecules in dry air are nitrogen (N_2). Another 21% of the molecules in dry air are oxygen (O_2). The final 1% of dry air is a mixture of other gases. For any gas, at a given temperature and pressure, the number of molecules present in a particular volume is constant – see ideal gas law. So when water molecules (vapor) are introduced into that volume of dry air, the number of air molecules in the volume must decrease by the same number, if the temperature and pressure remain constant. (The addition of water molecules, or any other molecules, to a gas, without removal of an equal number of other molecules, will necessarily require a change in temperature, pressure, or total volume; that is, a change in *at least* one of these three parameters. If temperature and pressure remain constant, the volume increases, and the dry air molecules that were displaced will initially move out into the additional volume, after which the mixture will eventually become uniform through diffusion.) Hence the mass per unit volume of the gas—its density—decreases. Isaac

Newton discovered this phenomenon and wrote about it in his book *Opticks*.

Effects

Animals and Plants

Humidity is one of the fundamental abiotic factors that defines any habitat, and is a determinant of which animals and plants can thrive in a given environment. The human body dissipates heat by a perspiration and evaporation. Heat convection to the surrounding air, and thermal radiation are the primary modes of heat transport from the body. Under conditions of high relative humidity, the rate of evaporation of sweat from the skin decreases. Also, if the atmosphere is as warm as or warmer than the skin during times of high relative humidity, blood brought to the body surface cannot dissipate heat by conduction to the air, and a condition called hyperpyrexia results. With so much blood going to the external surface of the body, relatively less goes to the active muscles, the brain, and other internal organs. Physical strength declines, and fatigue occurs sooner than it would otherwise. Alertness and mental capacity also may be affected, resulting in *heat stroke* or hyperthermia.

Human Comfort

Humans are sensitive to humid air because the human body uses evaporative cooling as the primary mechanism to regulate temperature. Under humid conditions, the *rate* at which perspiration evaporates on the skin is lower than it would be under arid conditions. Because humans perceive the rate of heat transfer from the body rather than temperature itself, we feel warmer when the relative humidity is high than when it is low.

Some people experience difficulty breathing in high humidity environments. Some cases may possibly be related to respiratory conditions such as asthma, while others may be the product of anxiety. Sufferers will often hyperventilate in response, causing sensations of numbness, faintness, and loss of concentration, among others. Air conditioning reduces discomfort in the summer not only by reducing temperature, but also by reducing relative humidity. In winter, heating cold outdoor air can decrease relative humidity levels indoor to below 30%, leading to discomfort such as dry skin and excessive thirst.

Electronics

Many electronic devices have humidity specifications, for example, 5% to 95%. At the top end of the range, moisture may increase the conductivity of permeable insulators leading to malfunction. Too low relative humidity may make materials brittle. A particular danger to electronic items, regardless of the stated operating humidity range, is condensation. When an electronic item is moved from a cold place (e.g., garage, car, shed, an air conditioned space in the tropics) to a warm humid place (house, outside tropics), condensation may coat circuit boards and other insulators, leading to short circuit inside the equipment. Such short circuits may cause substantial permanent damage if the equipment is powered on before the condensation has evaporated.

A similar condensation effect can often be observed when a person wearing glasses comes in from the cold. It is advisable to allow electronic equipment to acclimatise for several hours, after being brought in from the cold, before powering on. Some electronic devices can detect such a change and indicate, when plugged in and usually with a small droplet symbol, that they cannot be used until the risk from condensation has passed. In situations where time is critical, increasing air flow through the device's internals when, such as removing the side panel from a PC case and directing a fan to blow into the case will reduce significantly the time needed to acclimatise to the new environment.

On the opposite, very low relative humidity level favours the buildup of static electricity, which may result in spontaneous shutdown of computers when discharges occur. Apart from spurious erratic function, electrostatic discharges can cause dielectric breakdown in solid state devices, resulting in irreversible damage. Data centres often monitor relative humidity levels for these reasons.

Building Construction

Traditional building designs typically had weak insulation, and it allowed air moisture to flow freely between the interior and exterior. The energy-efficient, heavily-sealed architecture introduced in the 20th century also sealed off the movement of moisture, and this has resulted in a secondary problem of condensation forming in and around walls, which encourages the development of mold and mildew. Additionally, buildings with foundations not properly sealed will allow water to flow through the walls due to capillary action of pores found in masonry products. Solutions for energy-efficient buildings that avoid condensation are a current topic of architecture.

Ice Cap Climate

An ice cap climate is a polar climate where the temperature never or almost never exceeds 0 °C (32 °F). The climate covers the areas around the poles, such as Antarctica and Greenland, as well as the highest mountaintops. Such areas are covered by a permanent layer of ice and have no vegetation, but they may have animal life, that usually feeds from the oceans. Due to their high latitudes, icecap climates experience 24 hours of sun in the summer and no sunshine in winter, the midnight sun and polar night.

Description

Under the Köppen climate classification, the ice cap climate is denoted as EF. Ice caps are defined as a climate with no months above 0 °C (32 °F). Such areas are found around the north and south pole, and on the top of the highest mountains. Since the temperature never exceeds the melting point of water, any snow or ice that accumulates remains there permanently, over time forming a large ice sheet. The ice cap climate is distinct from the tundra climate, or *ET*. A tundra climate has a summer season with temperatures consistently above freezing for several months. This summer is enough to melt the winter ice cover, which prevents the formation of ice sheets. Because of this, tundras have vegetation, while ice caps do not. Ice cap climate is the world's coldest climate, and includes the coldest places on Earth. Vostok, Antarctica is the coldest place in the world, having recorded a temperature of –89.2 °C (–128.6 °F).

Ice Sheets

The constant freezing temperatures cause the formation of large ice sheets in ice cap climates. These ice sheets, however, are not static, but slowly move off the continents into the surrounding waters. New snow and ice accumulation then replaces the ice that is lost. Precipitation is nearly non-existent in ice cap climates. It is never warm enough for rain, and usually too cold to generate snow. However, wind can blow snow on to the ice sheets from nearby tundras.

The ice sheets are often miles thick. Much of the land located under the ice sheets is actually below sea level, and would be under the ocean if the ice is removed. However, it is the weight of the ice itself that forces this land below sea level. If the ice was removed, the land would rise back up in an effect called post-glacial rebound. This effect is creating new land in formerly ice cap areas such as Sweden. The extreme pressure exerted by the ice allows for the formation of liquid water at low temperatures that would otherwise result in ice, while the ice sheet itself insulates liquid water from the cold above. The causes the formation of sub-glacial lakes, the largest being Lake Vostok in Antarctica.

Geologic History

Icecap climates have only existed in ice ages. There have been at least five such ice ages in the Earth's past. Outside these ages, the Earth seems to have been ice-free even in high latitudes. By the geologic definition, Earth is currently in an ice age, in that the planet has permanent ice caps. Factors that cause ice ages include changes to the atmosphere, the arrangement of continents, the energy received from the sun, volcanos, and meteor impacts.

The current era is believed to be the only time in Earth's history with ice caps at both poles. The Antarctic ice cap was formed after Antarctica split from South America, allowing the formation of the Antarctic Circumpolar Current. The Arctic ice cap was partially caused by the Azolla event, where a large number a ferns in the ocean died, sank, and never decayed, which trapped carbon dioxide beneath the ocean. There is a hypothesis that around 650 million years ago, the entire planet was frozen, called Snowball Earth. Essentially the entire planet had an ice cap climate. However, this theory is disputed, and even proponents suggest there was an area of periodic melting near the equator.

Locations

The two major areas with ice cap climates are Antarctica and Greenland. Some of the most northern extremes of Canada also have ice cap climates. In addition, a large portion of the Arctic Ocean near the North Pole remains frozen year round, effectively making it an icecap climate.

North Pole

The Arctic Ocean is located over the North Pole. As a result, the northern polar ice cap is the frozen ocean. The only large landmass to have an icecap climate is Greenland, but several smaller islands near the Arctic Ocean also have permanent ice caps. Ice cap climates aren't nearly as common on land at the North Pole as in Antarctica. This is because the Arctic Ocean moderates the temperatures of the surrounding land, making the extreme cold seen in Antarctica impossible. In fact, the coldest places in the northern hemisphere are in subarctic climates in Siberia, such as Verkhoyansk, which are much farther inland and lack the ocean's moderating effect.

South Pole

The continent of Antarctica is centered around the South Pole. Antarctica is surrounded on all sides by the Southern Ocean. The

Southern Ocean circles the entire planet at its latitude. As a result, high-speed winds circle around Antarctica, preventing warmer air from temperate zones from reaching the continent. While Antarctica does have some small areas of Tundra on the northern fringes, the vast majority of the continent is extremely cold and permanently frozen. Because it is climactically isolated from the rest of the Earth, the continent has extreme cold not seen anywhere else, and weather systems rarely penetrate into the continent.

Extraterrestrial

Mars, like Earth, has ice caps at its poles. In addition to frozen water, Mars ice caps also have frozen carbon dioxide, commonly known as dry ice. In addition, Mars has seasons similar to Earth. The North Pole of Mars has a permanent water ice cap and a winter-only dry ice cap. On the south pole, both the water and carbon dioxide are permanently frozen. Europa, a moon of Jupiter, is covered with a permanent sheet of water ice, similar to Antarctica. Scientists theorize that it may have sub-glacial lakes similar to those seen in Antarctica.

Life

There is very little life in ice cap climates. Vegetation can not grow on ice, and is non-existent. However, ice caps do have significant animal life. Most of this life feeds on life on the surrounding oceans. Well known examples are polar bears and penguins. The interior of continents are devoid of animal life. Antarctica has several subglacial lakes underneath its ice sheet. Scientists have theorized that there may be life forms living in these lakes. In the summer of 2011-2012, Russian scientists drilled an ice core into Lake Vostok in Antarctica, but the core has not yet been analyzed. Scientists are particularly worried about accidentally contaminating the subglacial lakes with life forms from outside.

Ice-albedo Feedback

Ice-albedo feedback (or snow-albedo feedback) is a positive feedback climate process where a change in the area of snow-covered land,ice caps, glaciers or sea ice alters the albedo. This change in albedo acts to reinforce the initial alteration in ice area. Cooling tends to increase ice cover and hence the albedo, reducing the amount of solar energy absorbed and leading to more cooling. Conversely, warming tends to decrease ice cover and hence the albedo, increasing the amount of solar energy absorbed, leading to more warming.

The effect also applies on the small scale to snow-covered surfaces. A small amount of snow melt exposes darker ground which absorbs more radiation, leading to more snowmelt.

The effect has mostly been discussed in terms of the recent trend of declining Arctic sea ice. Internal feedback processes may also potentially occur, as land ice melts and causes eustatic sea level rise, and also potentially induces earthquakes as a result of isostatic rebound, which further acts to disrupt glaciers, ice shelves, etc.

Impact Winter

An impact winter is a period of prolonged cold weather caused by the impact on the Earth of a large asteroid or comet. If such an impact occurred on land or the floor of a shallow sea, it could cause large amounts of dust or ash to be thrown into the Earth's atmosphere, blocking the Sun's light and dramatically lowering the amount of sunlight reaching the Earth's surface. Impact winter is one of the mechanisms proposed for extinction events, such as the asteroid impact at Chicxulub in Mexico which supposedly led to the extinction of the dinosaurs.

Depending on the size of the object, and the location and angle at which it hits the earth, material can be thrown into the atmosphere by two mechanisms:

1. The impact could eject large amounts of regolith (and perhaps shattered bedrock) into the atmosphere
2. The impact could produce a global firestorm or strike a heavily forested area, throwing up large amounts of smoke and ash into the atmosphere.

The latter scenario is the more dangerous, as the lighter particles from the fire would take weeks or months to fall back to earth, and could be distributed by jet streams around the world, making the cooling a global event.

The mechanism of impact winter is very similar to that which occurs after nuclear war, leading to nuclear winter. Volcanoes also eject large amounts of opaque material into the higher parts of the atmosphere, with large explosions such as the 1991 explosion of Mount Pinatubo and the much larger 1783 Laki eruption, having measurable effects on the world's climate. The simultaneous eruption of a number of large volcanoes, a catastrophic volcanic winter scenario, is another proposed mechanism for extinction events.

Possible occurrences

- 536-540 : The cold period between 536 and 540 that caused crop failures and widespread starvation, could have been caused by the impact of a comet.

In Popular Culture

Impact winters (along with nuclear and volcanic winters) are often the subject of science fiction novels and short stories. In the episode "Impact Winter" of the popular television show *The West Wing*, NASA sights a large asteroid that could possibly collide with Earth. President Bartlet recounts the worst case scenario, saying "If the asteroid hits the forests of Russia, a shower of burning rock rains down on those woods and starts a fire that burns, that shrouds the hemisphere in a blanket of soot and ash that blocks out the sun for weeks. 'Impact winter', they call it."

The *Doctor Who* audio drama *Blood of the Daleks* is set on a human colony world which has suffered an asteroid strike and is undergoing an impact winter. In *Chrono Trigger*, the impact of Lavos in 65,000,000 BC caused an impact winter that apparently lasted for millions of years.

Marine Layer

A marine layer is an air mass which develops over the surface of a large body of water such as the ocean or large lake in the presence of a temperature inversion. The inversion itself is usually initiated by the cooling effect of the water on the surface layer of an otherwise warm air mass. As it cools, the surface air becomes denser than the warmer air above it, and thus becomes trapped below it. The layer may thicken through turbulence generated within the developing marine layer itself. It may also thicken if the warmer air above it is lifted by an approaching area of low pressure. The layer will also gradually increase its humidity by evaporation of the ocean or lake surface, as well as by the effect of cooling itself. Fog will form within a marine layer where the humidity is high enough and cooling sufficient to produce condensation. Stratus and stratocumulus will also form at the top of a marine layer in the presence of the same conditions there.

In the case of coastal California, the offshore marine layer is typically propelled inland by a pressure gradient which develops as a result of intense heating inland, blanketing coastal communities in cooler air which, if saturated, also contains fog. The fog lingers until the heat of the sun becomes strong enough to evaporate it, often lasting into the afternoon during the "May gray" or "June gloom" period. An approaching frontal system or trough can also drive the marine layer onshore. With Southern California's high concentration of military bases, a marine layer propelled inland can be colloquially described as "the marines coming inland".

A marine layer will disperse and break up in the presence of instability such as may be caused by the passage of frontal system or trough, or any upper air turbulence which reaches the surface. A marine layer can also be driven away by sufficiently strong winds.

It is not unusual to hear media weather reporters discuss the marine layer as synonymous with the fog or stratus it may contain, but this is erroneous. In fact, a marine layer can exist with virtually no cloudiness of any kind, although it usually does contain some. The marine layer is a medium within which clouds may form under the right conditions, not the layers of clouds themselves.

5

Western Disturbance

Western Disturbance is the term used in India, Pakistan, Bangladesh and Nepal to escribe an extratropical storm originating in the Mediterranean, that brings sudden winter rain and snow to the northwestern parts of the Indian subcontinent. This is a non-monsoonal precipitation pattern driven by the Westerlies. The moisture in these storms usually originates over the Mediterranean Sea and the Atlantic Ocean. Extratropical storms are a global, rather than a localized, phenomena with moisture usually carried in the upper atmosphere (unlike tropical storms where it is carried in the lower atmosphere). In the case of the subcontinent, moisture is sometimes shed as rain when the storm system encounters the Himalayas. Western Disturbances are important to the development of the Rabi crop in the northern subcontinent, which includes the locally important staple wheat.

Wind Rose

A wind rose is a graphic tool used by meteorologists to give a succinct view of how wind speed and direction are typically distributed at a particular location. Historically, wind roses were predecessors of the compass rose (found on maps), as there was no differentiation between a cardinal directionand the wind which blew from such a direction. Using a polar coordinate system of gridding, the frequency of winds over a long time period are plotted by wind direction, with colour bands showing wind ranges. The directions of the rose with the longest spoke show the wind direction with the greatest frequency.

History

Before the development of the compass rose, a wind rose was included on maps in order to let the reader know which directions the

8 major winds (and sometimes 8 half winds and 16 quarter winds) blew within the plan view. No differentiation was made between cardinal directions and the winds which blew from said direction. North was depicted with a fleur de lis, while east was shown as a Christian cross to indicate the direction of Jerusalemfrom Europe.

Use

Presented in a circular format, the modern wind rose shows the frequency of winds blowing *from* particular directions over a thirty year period. The length of each "spoke" around the circle is related to the frequency that the wind blows from a particular direction per unit time. Each concentric circle represents a different frequency, emanating from zero at the centre to increasing frequencies at the outer circles. A wind rose plot may contain additional information, in that each spoke is broken down into colour-coded bands that show wind speed ranges. Wind roses typically use 16 cardinal directions, such as north (N), NNE, NE, etc., although they may be subdivided into as many as 32 directions. In terms of angle measurement in degrees, North corresponds to 0°/360°, East to 90°, South to 180° and West to 270°. Compiling a wind rose is one of the preliminary steps taken in constructing airport runways, as aircraft perform their best take offs and landings pointing into the wind.

Younger Dryas

The Younger Dryas stadial, also referred to as the *Big Freeze*, was a geologically brief (1,300 ± 70 years) period of cold climatic conditions and drought which occurred between approximately 12,800 and 11,500 years BP (before present). The Younger Dryas stadial is thought to have been caused by the collapse of the North American ice sheets, although rival theories have been proposed. It followed the Bølling-Allerød interstadial (warm period) at the end of the Pleistocene and preceded the preboreal of the early Holocene. It is named after an indicator genus, the alpine-tundra wildflower *Dryas octopetala*. In Ireland, the period has been known as the Nahanagan Stadial, while in the United Kingdom it has been called the Loch Lomond Stadial and most recently Greenland Stadial 1 (GS1). The Younger Dryas (GS1) is also a Blytt-Sernander climate period detected from layers in north European bog peat. The Dryas stadials were cold periods which interrupted the warming trend since the Last Glacial Maximum 20,000 years ago. The Older Dryasoccurred approximately 1,000 years before the Younger Dryas and lasted about 300 years. The Oldest Dryas is dated between approximately 18,000 and 15,000 BP.

Abrupt Climate Change

The Younger Dryas saw a rapid return to glacial conditions in the higher latitudes of the Northern Hemisphere between 12.9–11.5 ka BP, in sharp contrast to the warming of the preceding interstadial deglaciation. It has been believed that the transitions each occurred over a period of a decade or so, but the onset may have been faster. Thermally fractionated nitrogen and argon isotope data from Greenland ice core GISP2 indicate that the summit of Greenland was approximately 15 °C (27.0 °F) colder during the Younger Dryas than today. In the UK, coleopteran fossil evidence (from beetles) suggests that mean annual temperature dropped to approximately 5 °C (41 °F), and periglacial conditions prevailed in lowland areas, while icefields and glaciers formed in upland areas. Nothing of the size, extent, or rapidity of this period of abrupt climate change has been experienced since.

Global Effects

In western Europe and Greenland, the Younger Dryas is a well-defined synchronous cool period. But cooling in the tropical North Atlantic may have preceded this by a few hundred years; South America shows a less well defined initiation but a sharp termination. The Antarctic Cold Reversal appears to have started a thousand years before the Younger Dryas, and has no clearly defined start or end; Huybers has argued that there is fair confidence in the absence of the Younger Dryas in Antarctica, New Zealand and parts of Oceania. Timing of the tropical counterpart to the Younger Dryas – the Deglaciation Climate Reversal (DCR) – is difficult to establish as low latitude ice core records generally lack independent dating over this interval. An example of this is the Sajama ice core (Bolivia), for which the timing of the DCR has been pinned to that of the GISP2 ice core record (central Greenland). Climatic change in the central Andes during the DCR, however, was significant and characterized by a shift to much wetter, and likely colder, conditions. The magnitude and abruptness of these changes would suggest that low latitude climate did not respond passively during the YD/DCR.

In western North America it is likely that the effects of the Younger Dryas were less intense than in Europe; however, evidence of glacial re-advance indicates Younger Dryas cooling occurred in the Pacific Northwest.

Other features seen include:

- Replacement of forest in Scandinavia with glacial tundra (which is the habitat of the plant *Dryas octopetala*).

- Glaciation or increased snow in mountain ranges around the world.
- Formation of solifluction layers and loess deposits in Northern Europe.
- More dust in the atmosphere, originating from deserts in Asia.
- Drought in the Levant, perhaps motivating the Natufian culture to develop agriculture.
- The Huelmo/Mascardi Cold Reversal in the Southern Hemisphere ended at the same time.
- Decline of the Clovis Culture and extinction of animal species in North America.

Causes

The prevailing theory holds that the Younger Dryas was caused by a significant reduction or shutdown of the North Atlantic "Conveyor", which circulates warm tropical waters northward, in response to a sudden influx of fresh water from Lake Agassiz and deglaciation in North America; however, geological evidence for such an event is thus far lacking.

The global climate would then have become locked into the new state until freezing removed the fresh water "lid" from the north Atlantic Ocean.

A recent alternative theory suggests instead that the jet stream shifted northward in response to the changing topographic forcing of the melting North American ice sheet, bringing more rain to the North Atlantic which freshened the ocean surface enough to slow the thermohaline circulation.

Previous glacial terminations probably did not have Younger Dryas-like events, suggesting that its cause has a random component. Nevertheless, there is evidence that some previous glacial terminations had post glacial cooling periods somewhat similar to the Younger Dryas.

Impact Event

A hypothesized Younger Dryas impact event, presumed to have occurred in North America around 12.9 ka BP, has been proposed as the mechanism to have initiated the Younger Dryas cooling, but reviews of the evidence for an impact event has shown that none of the original impact signatures that were attributed to the YD have been corroborated by independent tests; although some evidence, from one lake in central Mexico, may support the hypothesized impact event.

Volcanoes

Although there may be several causes of the Younger Dryas, volcanic activity is considered a plausible causative event. The Laacher See volcano in Germany was of sufficient size, VEI6, with over 10 km (2.4 cu mi) tephra ejected, to have caused significant temperature changes in the northern hemisphere. Laacher See tephra is found throughout the Younger Dryas boundary layer.

End of the Climate Period

Measurements of oxygen isotopes from the GISP2 ice core suggest the ending of the Younger Dryas took place over just 40 – 50 years in three discrete steps, each lasting five years. Other proxy data, such as dust concentration, and snow accumulation, suggest an even more rapid transition, requiring about a 7 °C (12.60 °F) warming in just a few years. Total warming was 10 ± 4 °C (18 ± 7 °F). The end of the Younger Dryas has been dated to around 11.55 ka BP, occurring at 10 ka BP (radiocarbon year), a "radiocarbon plateau") by a variety of methods, with mostly consistent results:

11.50 ± 0.05	ka BP	— GRIP ice core, Greenland
11.53 + 0.04- 0.06	ka BP	— Kråkenes Lake, western Norway
11.57	ka BP	— Cariaco Basin core, Venezuela
11.57	ka BP	— German oak/pine dendrochronology
11.64 ± 0.28	ka BP	— GISP2 ice core, Greenland

Effect on agriculture

The Younger Dryas is often linked to the adoption of agriculture in the Levant. It is argued that the cold and dry Younger Dryas lowered the carrying capacity of the area and forced the sedentary Early Natufian population into a more mobile subsistence pattern. Further climatic deterioration is thought to have brought about cereal cultivation. While there exists relative consensus regarding the role of the Younger Dryas in the changing subsistence patterns during the Natufian, its connection to the beginning of agriculture at the end of the period is still being debated. See the Neolithic Revolution, when hunter gatherers turned to farming.

Cultural references

The failure of North Atlantic thermohaline circulation is used to explain rapid climate change in some science fiction writings as early as Stanley G. Weinbaum's 1937 short story "Shifting Seas" where the author described the freezing of Europe after the Gulf Stream was disrupted, and more recently in Kim Stanley Robinson's novels, particularly *Fifty Degrees Below*. It also underpinned the 1999 book, *The Coming Global Superstorm*. Likewise, the idea of rapid climate change caused by disruption of North Atlantic ocean currents creates the setting for 2004 apocalyptic science-fiction film *The Day After Tomorrow*. Similar sudden cooling events have featured in other novels, such as John Christopher's *The World in Winter*, though not always with the same explicit links to the Younger Dryas event as is the case of Robinson's work.

Boreal ecosystem

The term *boreal* is usually applied to ecosystems localized in subarctic (Northern hemisphere) and subantarctic (Southern hemisphere) zones, although Austral is also used for the latter.

The ecosystems that lie immediately to the south (in the Northern hemisphere) or to the north (in the Southern hemisphere) of boreal zones are often called hemiboreal.

A "boreal forest", also known as the Taiga, is a forest ecosystem that can survive in northern, specifically subarctic, regions. The Koppen symbols of boreal ecosystems are Dfc, Dwc, Dfd and Dwd.

Taiga

Taiga also known as the boreal forest, is abiome characterized by coniferous forests. Taiga is the world's largest terrestrial biome. In North America it covers most of inland Canada and Alaska as well as parts of the extreme northern continental United States and is known as the Northwoods. It also covers most of Sweden, Finland, inland Norway, much of Russia (especially Siberia), northern Kazakhstan, northern Mongolia, and northern Japan (on the island of Hokkaidô).

The term "boreal forest" is sometimes, particularly in Canada, used to refer to the more southerly part of the biome, while the term taiga is often used to describe the more barren areas of the northernmost part of the taiga approaching the tree line.

Climate and geography

Taiga is the world's *largest* land biome, and makes up 29% of the world's forest cover; the largest areas are located in Russia and Canada. The taiga is the terrestrial biome with the lowest annual average temperatures after the tundra and permanent ice caps. Extreme winter minimums in the northern taiga are typically lower than those of the tundra. The lowest reliably recorded temperatures in the Northern

Hemisphere were recorded in the taiga of northeastern Russia. The taiga or boreal forest has asubarctic climate with very large temperature range between seasons, but the long and cold winter is the dominant feature. This climate is classified as *Dfc*, *Dwc*, *Dsc*, *Dfd* and *Dwd* in the Köppen climate classification scheme, meaning that the short summer (24-hr average 10 °C or more) lasts 1–3 months and always less than 4 months. There are also some much smaller areas grading towards the oceanic *Cfc* climate with milder winters, whilst the extreme south and (in Eurasia) west of the taiga reaches into humid continental climates (*Dfb*, *Dwb*) with longer summers. The mean annual temperature generally varies from -5 °C to 5 °C, but there are taiga areas in eastern Siberia and interior Alaska-Yukon where the mean annual reaches down to -10 °C. According to some sources, the boreal forest grades into a temperate mixed forest when mean annual temperature reaches about 3 °C. Discontinuous permafrost is found in areas with mean annual temperature below 0 °C, whilst in the *Dfd* and *Dwd* climate zones continuous permafrost occurs and restricts growth to very shallow-rooted trees like Siberian larch. The winters, with average temperatures below freezing, last five to seven months. Temperatures vary from –54 °C to 30 °C (-65 °F to 86 °F) throughout the whole year. The summers, while short, are generally warm and humid. In much of the taiga, -20 °C would be a typical winter day temperature and 18 °C an average summer day.

The growing season, when the vegetation in the taiga comes alive, is usually slightly longer than the climatic definition of summer as the plants of the boreal biome have a lower threshold to trigger growth. In Canada, Scandinavia and Finland, the growing season is often estimated by using the period of the year when the 24-hr average temperature is 5 °C or more. For the Taiga Plains in Canada, growing season varies from 80 to 150 days, and in the Taiga Shield from 100 to 140 days. Some sources claim 130 days growing season as typical for the taiga. Other sources mention that 50–100 frost-free days are characteristic. Data for locations in southwest Yukon gives 80–120 frost-free days. The closed canopy boreal forest in Kenozyorsky National Park near Plesetsk, Arkhangelsk Province, Russia, on average has 108 frost-free days. The longest growing season is found in the smaller areas with oceanic influences; in coastal areas of Scandinavia and Finland, the growing season of the closed boreal forest can be 145–180 days. The shortest growing season is found at the northern taiga–tundra ecotone, where the northern taiga forest no longer can grow and the tundra dominates the landscape when the growing season is down to 50–70 days, and the 24-hr average of the warmest month of the year usually is 10 °C or less. High latitudes mean that the sun does not rise far above the horizon, and less solar energy is received than further south. But the high latitude also ensures very long summer days, as the sun stays above the horizon nearly 20 hours each day, with only around 6 hours of daylight occurring in the dark winters, depending on latitude. The areas of the taiga inside the Arctic circle have midnight sun in mid-summer and polar night in mid-winter. The taiga experiences relatively low precipitation throughout the year (generally 200–750 mm annually, 1,000 mm in some areas), primarily as rain during the summer months, but also as fogand snow. This fog, especially predominant in low-lying areas during and after the thawing of frozen Arctic seas, means that sunshine is not abundant in the taiga even during the long summer days. As evaporation is consequently low for most of the year, precipitation exceeds evaporation, and is sufficient to sustain the dense vegetation growth. Snow may remain on the ground for as long as nine months in the northernmost extensions of the taiga ecozone.

In general, taiga grows to the south of the 10 °C July isotherm, but occasionally as far north as the 9 °C July isotherm. The southern limit is more variable, depending on rainfall; taiga may be replaced by forest steppe south of the 15 °C July isotherm where rainfall is very low, but more typically extends south to the 18 °C July isotherm, and locally where rainfall is higher (notably in eastern Siberia and adjacent Outer Manchuria) south to the 20 °C July isotherm. In these warmer areas the taiga has higher species diversity, with more warmth-loving species such as Korean Pine, Jezo Spruce, and Manchurian Fir, and merges gradually into mixed temperate forest or, more locally (on the Pacific Ocean coasts of North America and Asia), into coniferous temperate rainforests.

The area currently classified as taiga in Europe and North America (except Alaska) was recently glaciated. As the glaciers receded they leftdepressions in the topography that have since filled with water, creating lakes and bogs (especially muskeg soil) found throughout the taiga.

Soils

Taiga soil tends to be young and poor in nutrients. It lacks the deep, organically enriched profile present in temperate deciduous forests. The thinness of the soil is due largely to the cold, which hinders the development of soil and the ease with which plants can use its nutrients. Fallen leaves and moss can remain on the forest floor for a long time in the cool, moist climate, which limits their organic contribution to the soil; acids from evergreen needles further leach the soil, creating spodosol, also known as podzol. Since the soil is acidic due to the falling pine needles, the forest floor has only lichens and some mosses growing on it. In clearings in the forest and in areas with more boreal deciduous trees, there are more herbs and berries growing. Diversity of soil organisms in the boreal forest is high, comparable to the tropical rainforest.

Flora

Since North America and Asia used to be connected by the Bering land bridge, a number of animal and plantspecies (more animals than plants) were able to colonize both continents and are distributed throughout the taiga biome. Others differ regionally, typically with each genus having several distinct species, each occupying different regions of the taiga. Taigas also have some small-leaved deciduous trees like birch, alder, willow, and poplar; mostly in areas escaping the most extreme winter cold.

However, the Dahurian Larch tolerates the coldest winters in the northern hemisphere in eastern Siberia. The very southernmost parts of the taiga may have trees such as oak, maple, elm, and tilia scattered among the conifers, and there is usually a gradual transition into a temperate mixed forest, such as the Eastern forest-boreal transition of eastern Canada. In the interior of the continents with the driest climate, the boreal forests might grade into temperate grassland.

There are two major types of taiga. The southern part is the closed canopy forest, consisting of many closely spaced trees with mossy ground cover. In clearings in the forest, shrubs and wildflowers are common, such as the fireweed. The other type is the lichen woodland or sparse taiga, with trees that are farther-spaced and lichen ground cover; the latter is common in the northernmost taiga. In the northernmost taiga the forest cover is not only more sparse, but often stunted in growth form; moreover, ice pruned asymmetric Black Spruce (in North America) are often seen, with diminished foliage on the windward side. In Canada, Scandinavia and Finland, the boreal forest is usually divided into three subzones: The high boreal (north boreal) or taiga zone; the middle boreal (closed forest); and the southern boreal, a closed canopy boreal forest with some scattered temperate deciduous trees among the conifers, such as maple, elm and oak. This southern boreal forest experiences the longest and warmest growing season of the biome, and in some regions (including Scandinavia, Finland and western Russia) this subzone is commonly used for agricultural purposes. The boreal forest is home to many types of berries; some are confined to the southern and middle closed boreal forest (such as raspberry), others grow in most areas of the taiga (such as cranberry and cloudberry), and some can grow in both the taiga and the low arctic (southern part of) tundra (such as bilberry and lingonberry).

The forests of the taiga are largely coniferous, dominated by larch, spruce, fir, and pine. The woodland mix varies according to geography and climate so for example the Eastern Canadian forests ecoregion of the higher elevations of the Laurentian Mountains and the northern Appalachian Mountains in Canada is dominated by balsam fir *Abies balsamea*, while further north the Eastern Canadian Shield taiga of northern Quebec and Labrador is notably black spruce *Picea mariana* and tamarack larch *Larix laricina*.

Evergreen species in the taiga (spruce, fir, and pine) have a number of adaptations specifically for survival in harsh taiga winters, although larch, the most cold-tolerant of all trees, is deciduous. Taiga trees tend to have shallow roots to take advantage of the thin soils, while many of them seasonally alter their biochemistry to make them more resistant to freezing, called "hardening". The narrow conical shape of northern conifers, and their downward-drooping limbs, also help them shed snow. Because the sun is low in the horizon for most of the year, it is difficult for plants to generate energy from photosynthesis. Pine, spruce and fir do not lose their leaves seasonally and are able to photosynthesize with their older leaves in late winter and spring when light is good but temperatures are still too low for new growth to commence. The adaptation of evergreen needles limits the water lost due to transpiration and their dark green colour increases their absorption of sunlight. Although precipitation is not a limiting factor, the ground freezes during the winter months and plant roots are unable to absorb water, so desiccation can be a severe problem in late winter for evergreens.

Although the taiga is dominated by coniferous forests, some broadleaf trees also occur, notably birch, aspen, willow, and rowan. Many small erherbaceous plants grow closer to the ground. Periodic stand-replacing wildfires (with return times of between 20–200 years) clear out the tree canopies, allowing sunlight to invigorate new growth on the forest floor. For some species, wildfires are a necessary part of the life cycle in the taiga; some, e.g. Jack Pine have cones which only open to release their seed after a fire, dispersing their seeds onto the newly cleared ground. Grasses grow wherever they can find a patch of sun, and mosses and lichens thrive on the damp ground and on the sides of tree trunks. In comparison with other biomes, however, the taiga has low biological diversity.

Coniferous trees are the dominant plants of the taiga biome. A very few species in four main genera are found: the evergreen spruce, fir, and pine, and the deciduous larch. In North America, one or two species of fir and one or two species of spruce are dominant. Across Scandinavia and western Russia, the Scots pine is a common component of the taiga, while taiga of the Russian Far East and Mongolia is dominated bylarch.

Fauna

The boreal forest, or taiga, supports a large range of animals. Canada's boreal forest includes 85 species of mammals, 130 species of fish, and an estimated 32,000 species of insects. Insects play a critical role as pollinators, decomposers, and as a part of the food web. Many nesting birds rely on them for food. The cold winters and short summers make the taiga a challenging biome for reptiles and amphibians, which depend on environmental conditions to regulate their body temperatures, and there are only a few species in the boreal forest. Some hibernate underground in winter.

The taiga is home to a number of large herbivorous mammals, such as moose and reindeer/caribou. Some areas of the more southern closed boreal forest also have populations of other deer species such as the elk (wapiti) and roe deer. There is also a range of rodent species including beaver, squirrel, mountain hare, snowshoe hare, and vole. These species have evolved to survive the harsh winters in their native ranges. Some larger mammals, such as bears, eat heartily during the summer in order to gain weight, and then go into hibernation during the winter. Other animals have adapted layers of fur or feathers to insulate them from the cold. A number of wildlife species threatened or endangered with extinction can be found in the Canadian boreal forest, including woodland caribou, American black bear, grizzly bear, wood bison and wolverine. Habitat loss, mainly due to logging, is the primary cause of decline for these species.

Due to the climate, carnivorous diets are an inefficient means of obtaining energy; energy is limited, and most energy is lost between trophic levels. Predatory birds (owls and eagles) and other smaller carnivores, including foxes and weasels, feed on the rodents. Larger carnivores, such as lynx and wolves, prey on the larger animals. Omnivores, such as bears andraccoons are fairly common, sometimes picking through human garbage.

More than 300 species of birds have their nesting grounds in the taiga. Siberian Thrush, White-throated Sparrow, and Black-throated Green Warbler migrate to this habitat to take advantage of the long summer days and abundance of insects found around the numerous bogs and lakes. Of the 300 species of birds that summer in the taiga only 30 stay for the winter. These are either carrion-feeding or large raptors that can take live mammal prey, including Golden Eagle, Rough-legged Buzzard (also known as the Rough-legged Hawk), and Raven, or else seed-eating birds, including several species of grouse and crossbills.

Threats

Human Activities

Large areas of Siberia's taiga have been harvested for lumber since the collapse of the Soviet Union. In Canada, eight percent of the boreal forest is protected from development, the provincial government allows forest management to occur on Crown land under rigorous constraints. The main forestry practice in the boreal forest of Canada is clearcutting, which involves cutting down most of the trees in a given area, then replanting the forest as a monocrop (one species of tree) the following season. Industry officials claim that this process emulates the natural effects of a forest fire, which they claim clearcutting suppresses, protecting infrastructure, communities and roads. However, from an ecological perspective, this is a falsehood, for several reasons, including: a) Removing most of the trees in a given area is usually done using large machines which disrupt the soil greatly, and the dramatic diminution of ground cover permits large-scale erosion and avalanches, which further damage the habitat and sometimes endangers infrastructure, roads, and communities. b) Clearcutting removes most of the biomass from an area, and the various macro and micro-nutrients it contains. This sudden decrease in nutrients in an area contrasts with a forest fire, which returns most of the nutrients to the soil. c) Forest fires leave standing snags, and leave patches of unburned trees. This helps preserve structure and micro-habitats within the area, whereas clearcutting destroys most of these habitats. In the past, clearcuts upwards of 110 km^2 have been recorded in the Canadian boreal forest. However, today 80% of clearcuts are less than 260 hectares (2.6 square km). Some of the products from logged boreal forests include toilet paper, copy paper, newsprint, and lumber. More than 90% of boreal forest products from Canada are exported for consumption and processing in the United States. However with the recession and fewer US homes being built, that has changed. Some of the larger cities situated in this biome are Murmansk, Arkhangelsk, Yakutsk, Anchorage, Yellowknife, Tromsø, Luleå, and Oulu.

Most companies that harvest in Canadian forests are certified by an independent third party agency such as the Forest Stewardship Council (FSC), Sustainable Forests Initiative (SFI), or the Canadian Standards Association (CSA). While the certification process differs between these groups, all of them include forest stewardship, respect for aboriginal peoples, compliance with local, provincial or national environmental laws, forest worker safety, education and training, and other environmental, business, and social requirements. The prompt renewal of all harvest sites by planting or natural renewal is also required.

Climate Change

The zone of latitude occupied by the boreal forest has experienced some of the greatest temperature increases on Earth, especially during the last quarter of the twentieth century. Winter temperatures have increased more than summer temperatures. The number of days with extremely cold temperatures (e.g., "20 to –40 °C) has decreased irregularly but systematically in nearly all the boreal region, allowing better survival for tree-damaging insects. In summer, the daily low temperature has increased more than the daily high temperature. In Fairbanks, Alaska, the length of the frost-free season has increased from 60–90 days in the early twentieth century to about 120 days a century later. Summer warming has been shown to increase water stress and reduce tree growth in dry areas of the southern boreal forest in central Alaska, western Canada and portions of far eastern Russia. Precipitation is relatively abundant in Scandinavia, Finland, northwest Russia and eastern Canada, where a longer growth season (i.e. the period when sap flow is not impeded by frozen water) accelerate tree growth. As a consequence of this warming trend, the warmer parts of the boreal forests are susceptible to replacement by grassland, parkland or temperate forest.

In Siberia, the taiga is converting from predominantly needle-shedding larch trees to evergreen conifers in response to a warming climate. This is likely to further accelerate warming, as the evergreen trees will absorb more of the sun's rays. Given the vast size of the area, such a change has the potential to affect areas well outside of the region. In much of the boreal forest in Alaska, the growth of white spruce trees are stunted by unusually warm summers, while trees on some of the coldest fringes of the forest are experiencing faster growth than previously. Lack of moisture in the warmer summers are also stressing the birch trees of central Alaska.

Insects

Recent years have seen outbreaks of insect pests in forest-destroying plagues: the spruce-bark beetle (*Dendroctonus rufipennis*) in the Yukon Territory, Canada, and Alaska; theaspen-leaf miner; the larch sawfly; the spruce budworm (*Choristoneura fumiferana*); the spruce coneworm.

Protection

Many nations are taking direct steps to protect the ecology of the taiga by prohibiting logging, mining, oil and gas production, and other forms of development. In February 2010 the Canadian government established protection for 13,000 square kilometres of boreal forest by creating a new 10,700 square kilometre park reserve in the Mealy Mountains area of eastern Canada and a 3,000 square kilometre waterway provincial park that follows alongside the Eagle River from headwaters to sea.

The taiga stores enormous quantities of carbon, possibly more than the temperate and tropical forests combined, much of it in peatland.

Natural Disturbance

One of the biggest areas of research and a topic still full of unsolved questions is the recurring disturbance of fire and the role it plays in propagating the lichen woodland. The phenomenon of wildfire by lightning strike is the primary determinant of understory vegetation and because of this, it is considered to be predominate driving force behind community and ecosystem properties in the lichen woodland. The significance of fire is clearly evident when one considers that understory vegetation influences tree seedling germination in the short term and decomposition of biomass and nutrient availability in the long term. The recurrent cycle of large, damaging fire occurs approximately every 70 to 100 years. Understanding the dynamics of this ecosystem is entangled with discovering the successional paths that the vegetation exhibits after a fire. Trees, shrubs and lichens all recover from fire induced damage through vegetative reproduction as well as invasion by propagules. Seeds that have fallen and become buried provide little help in re-establishment of a species. The reappearance of lichens is reasoned to occur because of varying conditions and light/nutrient availability in each different microstate. Several different studies have been done that have led to the formation of the theory that post-fire development can be propagated by any of four pathways: self replacement, species-dominance relay, species replacement, or gap-phase self replacement. Self replacement is simply the re-establishment of the pre-fire dominant species. Species-dominance relay is a sequential attempt of tree species to establish dominance in the canopy. Species replacement is when fires occur in sufficient frequency to interrupt species dominance relay. Gap-Phase Self-Replacement is the least common and so far has only been documented in Western Canada. It is a self replacement of the surviving species into the canopy gaps after a fire kills another species. The particular pathway taken after a fire disturbance depends on how the landscape is able to support trees as well as fire frequency. Fire frequency has a large role in shaping the original inception of the lower forest line of the lichen woodland taiga.

It has been hypothesized by Serge Payette that the Spruce-Moss forest ecosystem was changed into the lichen woodland biome due to the initiation of two compounded strong disturbances: large fire and the appearance and attack of the spruce budworm. The spruce budworm is a deadly insect to the spruce populations in the southern regions of the taiga. J.P. Jasinski confirmed this theory five years later stating "Their [lichen woodlands] persistence, along with their previous moss forest histories and current occurrence adjacent to closed moss forests, indicate that they are an alternative stable state to the spruce–moss forests".

Albedo

Albedo, or *reflection coefficient,* derived from Latin *albedo* "whiteness" (or reflected sunlight), in turn from *albus* "white", is thediffuse reflectivity or reflecting power of a surface. It is defined as the ratio of reflected radiation from the surface to incident radiation upon it. Being a dimensionless fraction, it may also be expressed as a percentage, and is measured on a scale from zero for no reflecting power of a perfectly black surface, to 1 for perfect reflection of a white surface.

Albedo depends on the frequency of the radiation. When quoted unqualified, it usually refers to some appropriate average across the spectrum of visible light. In general, the albedo depends on the directional distribution of incoming radiation. Exceptions are Lambertian surfaces, which scatter radiation in all directions according to a cosine function, so their albedo does not depend on the incident distribution. In practice, abidirectional reflectance distribution function (BRDF) may be required to characterize the scattering properties of a surface accurately, although the albedo is a very useful first approximation.

The albedo is an important concept in climatology and astronomy, as well as in calculating reflectivity of surfaces in LEED sustainable rating systems for buildings, computer graphics and computer vision. The average overall albedo of Earth, its *planetary albedo*, is 30 to 35%, because of the covering by clouds, but varies widely locally across the surface, depending on the geological and environmental features.

The term was introduced into optics by Johann Heinrich Lambert in his 1760 work *Photometria*.

Terrestrial Albedo

Albedos of typical materials in visible light range from up to 0.9 for fresh snow, to about 0.04 for charcoal, one of the darkest substances. Deeply shadowed cavities can achieve an effective albedo approaching the zero of a black body. When seen from a distance, the ocean surface has a low albedo, as do most forests, while desert areas have some of the highest albedos among landforms. Most land areas are in an albedo range of 0.1 to 0.4. The average albedo of the Earth is about 0.3. This is far higher than for the ocean primarily because of the contribution of clouds. Human activities have changed the albedo (via forest clearance and farming, for example) of various areas around the globe. However, quantification of this effect on the global scale is difficult.

The classic example of albedo effect is the snow-temperature feedback. If a snow-covered area warms and the snow melts, the albedo decreases, more sunlight is absorbed, and the temperature tends to increase. The converse is true: if snow forms, a cooling cycle happens. The intensity of the albedo effect depends on the size of the change in albedo and the amount of insolation; for this reason it can be potentially very large in the tropics. The Earth's surface albedo is regularly estimated via Earth observation satellite sensors such as NASA's MODISinstruments on board the Terra and Aqua satellites. As the total amount of reflected radiation cannot be directly measured by satellite, a mathematical model of the BRDF is used to translate a sample set of satellite reflectance measurements into estimates of directional-hemispherical reflectance and bi-hemispherical reflectance. (e. g.,.)

The Earth's average surface temperature due to its albedo and the greenhouse effect is currently about 15 °C. For the frozen (more reflective) planet the average temperature is below –40 °C (If only all continents being completely covered by glaciers - the mean temperature is about 0 °C). The simulation for (more absorptive) aquaplanet shows the average temperature close to 27 °C.

White-sky and black-sky albedo

It has been shown that for many applications involving terrestrial albedo, the albedo at a particular solar zenith angle θ_i can reasonably be approximated by the proportionate sum of two terms: the directional-hemispherical reflectance at that solar zenith angle, $\bar{\alpha}(\theta_i)$, and the bi-hemispherical reflectance, $\bar{\bar{\alpha}}$ the proportion concerned being defined as the proportion of diffuse illumination D.

Albedo α can then be given as:

$$\alpha = (1 - D)\bar{\alpha}(\theta_i) + D\bar{\bar{\alpha}}.$$

Directional-hemispherical reflectance is sometimes referred to as black-sky albedo and bi-hemispherical reflectance as white sky albedo. These terms are important because they allow the albedo to be calculated for any given illumination conditions from a knowledge of the intrinsic properties of the surface.

Astronomical Albedo

The albedos of planets, satellites and asteroids can be used to infer much about their properties. The study of albedos, their dependence on wavelength, lighting angle ("phase angle"), and variation in time comprises a major part of the astronomical field of photometry. For small and far objects that cannot be resolved by telescopes, much of what we know comes from the study of their albedos.

For example, the absolute albedo can indicate the surface ice content of outer solar system objects, the variation of albedo with phase angle gives information about regolith properties, while unusually high radar albedo is indicative of high metallic content in asteroids. Enceladus, a moon of Saturn, has one of the highest known albedos of any body in the Solar system, with 99% of EM radiation reflected.

Another notable high-albedo body is Eris, with an albedo of 0.96. Many small objects in the outer solar system and asteroid belt have low albedos down to about 0.05. A typical comet nucleus has an albedo of 0.04.Such a dark surface is thought to be indicative of a primitive and heavily space weathered surface containing some organic compounds.

The overall albedo of the Moon is around 0.12, but it is strongly directional and non-Lambertian, displaying also a strong opposition effect. While such reflectance properties are different from those of any terrestrial terrains, they are typical of the regolith surfaces of airless solar system bodies.

Two common albedos that are used in astronomy are the (V-band) geometric albedo (measuring brightness when illumination comes from directly behind the observer) and the Bond albedo (measuring total proportion of electromagnetic energy reflected). Their values can differ significantly, which is a common source of confusion.

In detailed studies, the directional reflectance properties of astronomical bodies are often expressed in terms of the five Hapke parameters which semi-empirically describe the variation of albedo with phase angle, including a characterization of the opposition effect of regolith surfaces.

The correlation between astronomical (geometric) albedo, absolute magnitude and diameter is:

$$A = \left(\frac{1329 \times 10^{-H/5}}{D} \right)^2,$$

where A is the astronomical albedo, D is the diameter in kilometres, and H is the absolute magnitude.

Examples of Terrestrial Albedo Effects

Illumination

Although the albedo-temperature effect is best known in colder, whiter regions on Earth, the maximum albedo is actually found in the tropics where year-round illumination is greater. The maximum is additionally in the northern hemisphere, varying between 3 and 12 degrees north. The minima are found in the subtropical regions of the north and south hemispheres, beyond which albedo increases without respect to illumination.

Small-scale Effects

Albedo works on a smaller scale, too. In sunlight, dark clothes absorb more heat and light-coloured clothes reflect it better, thus allowing some control over body temperature by exploiting the albedo effect of the colour of external clothing.

Trees

Because trees tend to have a low albedo, removing forests would tend to increase albedo and thereby could produce localized climate cooling (ignoring the lost evaporative cooling effect of trees). Cloud feedbacks further complicate the issue. In seasonally snow-covered zones, winter albedos of treeless areas are 10% to 50% higher than nearby forested areas because snow does not cover the trees as readily.

Deciduous trees have an albedo value of about 0.15 to 0.18 while coniferous trees have a value of about 0.09 to 0.15. Studies by the Hadley Centre have investigated the relative (generally warming) effect of albedo change and (cooling) effect of carbon sequestration on planting forests. They found that new forests in tropical and midlatitude areas tended to cool; new forests in high latitudes (e.g. Siberia) were neutral or perhaps warming.

Snow

Snow albedos can be as high as 0.9; this, however, is for the ideal example: fresh deep snow over a featureless landscape. Over Antarctica they average a little more than 0.8. If a marginally snow-covered area warms, snow tends to melt, lowering the albedo, and hence leading to more snowmelt (the ice-albedo positive feedback). Cryoconite, powdery windblown dust containing soot, sometimes reduces albedo on glaciers and ice sheets.

Water

Water reflects light very differently from typical terrestrial materials. The reflectivity of a water surface is calculated using the Fresnel equations. At the scale of the wavelength of light even wavy water is always smooth so the light is reflected in a locally specular manner (notdiffusely). The glint of light off water is a commonplace effect ofthis. At small angles of incident light, waviness results in reduced reflectivity because of the steepness of the reflectivity-vs.-incident-angle curve and a locally increased average incident angle. Although the reflectivity of water is very low at low and medium angles of incident light, it increases tremendously at high angles of incident light such as occur on the illuminated side of the Earth near the terminator (early morning, late afternoon and near the poles). However, as mentioned above, waviness causes an appreciable reduction. Since the light specularly reflected from water does not usually reach the viewer, water is usually considered to have a very low albedo in spite of its high reflectivity at high angles of incident light.

Note that white caps on waves look white (and have high albedo) because the water is foamed up, so there are many superimposed bubble surfaces which reflect, adding up their reflectivities. Fresh 'black' ice exhibits Fresnel reflection.

Clouds

Cloud albedo has substantial influence over atmospheric temperatures. Different types of clouds exhibit different reflectivity, theoretically ranging in albedo from a minimum of near 0 to a maximum approaching 0.8. "On any given day, about half of Earth is covered by clouds, which reflect more sunlight than land and water. Clouds keep Earth cool by reflecting sunlight, but they can also serve as blankets to trap warmth."

Albedo and climate in some areas are affected by artificial clouds, such as those created by the contrails of heavy commercial airliner traffic. A study following the burning of the Kuwaiti oil fields during Iraqi occupation showed that temperatures under the burning oil fires were as much as 10 °C colder than temperatures several miles away under clear skies.

Aerosol Effects

Aerosols (very fine particles/droplets in the atmosphere) have both direct and indirect effects on the Earth's radiative balance. The direct (albedo) effect is generally to cool the planet; the indirect effect (the particles act as cloud condensation nuclei and thereby change cloud properties) is less certain. As per the effects are:

- *Aerosol direct effect.* Aerosols directly scatter and absorb radiation. The scattering of radiation causes atmospheric cooling, whereas absorption can cause atmospheric warming.
- *Aerosol indirect effect.* Aerosols modify the properties of clouds through a subset of the aerosol population called cloud condensation nuclei. Increased nuclei concentrations lead to increased cloud droplet number concentrations, which in turn leads to increased cloud albedo, increased light scattering and radiative cooling (*first indirect effect*), but also leads to reduced precipitation efficiency and increased lifetime of the cloud (*second indirect effect*).

Black Carbon

Another albedo-related effect on the climate is from black carbon particles. The size of this effect is difficult to quantify: the Intergovernmental Panel on Climate Change estimates that the global mean radiative forcing for black carbon aerosols from fossil fuels is +0.2 W m^{-2}, with a range +0.1 to +0.4 W m^{-2}.

Other Types of Albedo

Single scattering albedo is used to define scattering of electromagnetic waves on small particles. It depends on properties of the material (refractive index); the size of the particle or particles; and the wavelength of the incoming radiation.

Infrared Window

The infrared atmospheric window is the overall dynamic property of the earth's atmosphere, taken as a whole at each place and occasion of interest, that lets some infrared radiation from the cloud tops and land-sea surface pass directly to space without intermediate absorption and re-emission, and thus without heating the atmosphere.

It cannot be defined simply as a part or set of parts of the electromagnetic spectrum, because the spectral composition of window radiation varies greatly with varying local environmental conditions, such as water vapour content and land-sea surface temperature, and because few or no parts of the spectrum are simply not absorbed at all, and because some of the diffuse radiation is passing nearly vertically upwards and some is passing nearly horizontally. A large gap in the absorption spectrum of water vapor, the main greenhouse gas, is most important in the dynamics of the window. Other gases, especially carbon dioxide and ozone, partly block transmission.

One should take care to distinguish between the atmospheric window and the spectral window. An atmospheric window is a dynamic property of the atmosphere, while the spectral window is a static characteristic of the electromagnetic radiative absorption spectra of many greenhouse gases, including water vapour. The atmospheric window tells what actually happens in the atmosphere, while the spectral window tells of one of the several abstract factors that potentially contribute to the actual concrete happenings in the atmosphere.

It is also important to distinguish between the terms window radiation and radiation of window wavelength (window wavelength radiation). Window radiation is radiation that actually passes through the atmospheric window. Non-window radiation is radiation that actually does not pass through the atmospheric window. Window wavelength radiation is radiation that, judging only from its wavelength, potentially might or might not, but is likely to pass through the atmospheric window.

Non-window wavelength radiation is radiation that, judging only from its wavelength, is unlikely to pass through the atmospheric window. The difference between window radiation and window wavelength radiation is that window radiation is an actual component of the radiation, determined by the full dynamics of the atmosphere, taking in all determining factors, while window wavelength radiation is merely theoretically potential, defined only by one factor, the wavelength.

The importance of the infrared atmospheric window in the atmospheric energy balance was discovered by George Simpson in 1928, based on G. Hettner's 1918 laboratory studies of the gap in the absorption spectrum of water vapor. In those days, computers were not available, and Simpson notes that he used approximations; he writes: "There is no hope of getting an exact solution; but by making suitable simplifying assumptions. . . ." Nowadays, accurate line-by-line computations are possible, and careful studies of the infrared atmospheric window have been published.

Kinetics of the Infrared Atmospheric Window

The infrared atmospheric window is a path from the land-sea surface of the earth to space. It separates two radiative components, window and non-window radiation, that are not of the kind that have kinetics suitable for description by the Beer-Lambert law. The window radiation and the non-window radiation from the land-sea surface are not defined in the terms that are necessary for the application of the Beer-Lambert Law. It would therefore be a logical and conceptual error to try to apply the Beer-Lambert Law either to window or non-window radiation considered separately.

The reason for this is that the window and non-window radiation have already been conditioned by the Beer-Lambert Law and the law cannot validly be re-applied to its own products. Logically, the Beer-Lambert Law applies to radiation of which the origin is known but the destination is unknown. Such is not the case for window and non-window radiation.

Logically, it is part of the definition of window radiation that its destination is known, namely that it is destined to go to space, and likewise, by definition the destination of non-window radiation is known to be entire absorption by the atmosphere. Thus it makes sense to state the precise spectral distribution and spatial, especially altitudinal, distribution of locations of absorption of non-window radiation in the atmosphere. But none of those locations can be beyond the atmosphere; by definition, non-window radiation has zero probability of escaping absorption by the atmosphere; all of the locations of absorption are within the atmosphere. Radiation that can be described by the Beer-Lambert Law can partly escape absorption by the medium of interest; the law tells just how much that part is. This is a deep conceptual point that distinguishes the kinetic description of window and non-window radiation from the kinetic description of the kind of radiation that is covered by the Beer-Lambert Law.

Non-window radiation is by definition absorbed by the atmosphere, and its energy is thereby transduced into kinetic energy of atmospheric molecules. That kinetic energy is then transferred according to the usual dynamics of atmospheric energy transfer. These kinetic principles for window and non-window radiation arise in the light of the definition of the atmospheric window as a dynamic property of the whole atmosphere, logically distinct from the electromagnetic spectral window.

Mechanisms in the Infrared Atmospheric Window

The infrared absorptions of the principal *natural* greenhouse gases are mostly in two ranges. At wavelengths longer than 14 μm (micrometres), gases such as CO_2 and CH_4 (along with less abundant hydrocarbons) absorb due to the presence of relatively long C-H and carbonyl bonds, as well as water (H_2O) vapor absorbing in rotation modes. The bonds of H_2O and NH_3absorb at wavelengths shorter than 8 μm. Except for the bonds in O_3, no bonds between carbon, hydrogen, oxygen and nitrogen atoms absorb in the interval between about 8 and 14 μm, though there is weaker continuum absorption in that interval.

Over the Atlas Mountains, interferometrically recorded spectra of outgoing longwave radiation show emission that has arisen from the land surface at a temperature of about 320 K and passed through the atmospheric window, and non-window emission that has arisen mainly from the troposphere at temperatures about 260 K.

Over the Ivory Coast, interferometrically recorded spectra of outgoing longwave radiation show emission that has arisen from the cloud tops at a temperature of about 265 K and passed through the atmospheric window, and non-window emission that has arisen mainly from the troposphere at temperatures about 240 K.

This means that, at the scarcely absorbed continuum of wavelengths (8 to 14 μm), the radiation emitted, by the earth's surface into a dry atmosphere, and by the cloud tops, mostly passes unabsorbed through the atmosphere, and is emitted directly to space; there is also partial window transmission in far infrared spectral lines between about 16 and 28 μm. Clouds are excellent emitters of infrared radiation. Window radiation from cloud tops arises at altitudes where the air temperature is low, but as seen from those altitudes, the water vapor content of the air above is much lower than that of the air at the land-sea surface. Moreover, the water vapour continuum absorptivity, molecule for molecule, decreases with pressure decrease. Thus water vapour above the clouds, besides being less concentrated, is also less absorptive than water vapour at lower altitudes. Consequently, the effective window as seen from the cloud-top altitudes is more open, with the result that the cloud tops are effectively strong sources of window radiation; that is to say, in effect the clouds obstruct the window only to a small degree.

Importance for Life

Without the infrared atmospheric window, the Earth would become much too warm to support life, and possibly so warm that it would lose its water as Venus did early in solar systemhistory. Thus, the existence of an atmospheric window is critical to Earth remaining a habitable planet.

Threats

In recent decades, the existence of the infrared atmospheric window has become threatened by the development of highly unreactive gases containing bonds between fluorine and eithercarbon or sulfur. The "stretching frequencies" of bonds between fluorine and other light nonmetals are such that strong absorption in the atmospheric window will always be characteristic of compounds containing such bonds. This absorption is strengthened because these bonds are highly polar due to the extreme electronegativity of the fluorine atom. Bonds to othe rhalogens also absorb in the atmospheric window, though much less strongly. Moreover, the unreactive nature of such compounds that makes them so valuable for many industrial purposes means that they are not removable in the natural circulation of the Earth's atmosphere. It is estimated, for instance, that perfluorocarbons (CF_4, C_2F_6, C_3F_8) can stay in the atmosphere for over fifty thousand years, a figure which may be an underestimate given the absence of natural sources of these gases.

This means that such compounds have an enormous global warming potential. One kilogram of sulfur hexafluoride will, for example, cause as much warming as 23 tonnes of carbon dioxide over 100 years. Perfluorocarbons are similar in this respect, and even carbon tetrachloride (CCl_4) has a global warming potential of 1800 compared to carbon dioxide.

Efforts to find substitutes for these compounds are still going on and remain highly problematic.

Cloud Feedback

Cloud feedback is the coupling between cloudiness and surface air temperature in which a change in radiative forcing perturbs the surface air temperature, leading to a change in clouds, which could then amplify or diminish the initial temperature perturbation. Global warming is expected to change the distribution and type of clouds. Seen from below, clouds emit infrared radiation back to the surface, and so exert a warming effect; seen from above, clouds reflect sunlight and emit infrared radiation to space, and so exert a cooling effect. Cloud representations vary among global climate models, and small changes in cloud cover have a large impact on the climate. Differences in planetary boundary layer cloud modelling schemes can lead to large differences in derived values of climate sensitivity.

A model that decreases boundary layer clouds in response to global warming has a climate sensitivity twice that of a model that does not include this feedback. However, satellite data show that cloud optical thickness actually increases with increasing temperature. Whether the net effect is warming or cooling depends on details such as the type and altitude of the cloud; details that are difficult to represent in climate models. In addition to how clouds themselves will respond to increased temperatures, there exist other feedbacks that will affect clouds properties and formation. The amount and vertical distribution of water vapor is closely linked to the formation of clouds. Ice crystals have been shown to largely influence the amount of water vapor. Water vapor in the subtropical upper troposphere has been linked to the convection of water vapor and ice. Changes in subtropical humidity could provide a negative feedback that decreases the amount of water vapor which would act to mediate global climate transitions. Changes in cloud cover are closely coupled with other feedback, including the water vapor feedback and ice-albedo feedback. Changing climate is expected to alter the relationship between cloud ice and supercooled cloud water, which in turn would influence the microphysics of the cloud which would result in changes in the radiative properties of the cloud. Climate models suggest that a warming will increase fractional cloudiness. More clouds cools the climate, resulting in a negative feedback. Increasing temperatures in the polar regions is expected in increase the amount of low-level clouds, whose stratification prevents the convection of moisture to upper levels. This feedback would partially cancel the increased surface warming due to the cloudiness.

Cloud Forcing

Cloud forcing (sometimes described as cloud radiative forcing) is, in meteorology, the difference between the radiation budget components for average cloud conditions and cloud-free conditions. Much of the interest in cloud forcing relates to its role as a feedback process in the present period of global warming.

All global climate models used for climate change projections include the effects of water vapor and cloud forcing. The models include the effects of clouds on both incoming (solar) and emitted (terrestrial) radiation. Clouds increase the global reflection of solar radiation from 15% to 30%, reducing the amount of solar radiation absorbed by the Earth by about 44 W/m^2. This cooling is offset somewhat by the greenhouse effect of clouds which reduces the outgoing longwave radiation by about 31 W/m^2. Thus the net cloud forcing of the radiation budget is a loss of about 13 W/m^2. If the clouds were removed with all else remaining the same, the Earth would gain this last amount in net radiation and begin to warm up. These numbers should not be confused with the usualradiative forcing concept, which is for the *change* in forcing related to climate change.

Without the inclusion of clouds, water vapor alone contributes 36% to 70% of the greenhouse effect on Earth. When water vapor and clouds are considered together, the contribution is 66% to 85%. The ranges come about because there are two ways to compute the influence of water vapor and clouds: the lower bounds are the reduction in the greenhouse effect if water vapor and clouds are *removed* from the atmosphere leaving all other greenhouse gases unchanged, while the upper bounds are the greenhouse effect introduced if water vapor and clouds are *added* to an atmosphere with no other greenhouse gases. The two values differ because of overlap in the absorption and emission by the various greenhouse gases. Trapping of the long-wave radiation due to the presence of clouds reduces the radiative forcing of the greenhouse gases compared to the clear-sky forcing. However, the magnitude of the effect due to clouds varies for different greenhouse gases. Relative to clear skies, clouds reduce the global mean radiative forcing due to CO_2 by about 15%, that due to CH_4 and N_2O by about 20%, and that due to the halocarbons by up to 30%. Clouds remain one of the largest uncertainties in future projections of climate change by global climate models, owing to the physical complexity of cloud processes and the small scale of individual clouds relative to the size of the model computational grid.

6

Global-warming Potential

Global-warming potential (GWP) is a relative measure of how much heat a greenhouse gas traps in the atmosphere. It compares the amount of heat trapped by a certain mass of thegas in question to the amount of heat trapped by a similar mass of carbon dioxide. A GWP is calculated over a specific time interval, commonly 20, 100 or 500 years.

GWP is expressed as a factor of carbon dioxide (whose GWP is standardized to 1). For example, the 20 year GWP of methane is 72, which means that if the same mass of methane and carbon dioxide were introduced into the atmosphere, that methane will trap 72 times more heat than the carbon dioxide over the next 20 years.

The substances subject to restrictions under the Kyoto protocol either are rapidly increasing their concentrations in Earth's atmosphere or have a large GWP.

The GWP depends on the following factors:

- the absorption of infrared radiation by a given species
- the spectral location of its absorbing wavelengths
- the atmospheric lifetime of the species

Thus, a high GWP correlates with a large infrared absorption and a long atmospheric lifetime. The dependence of GWP on the wavelength of absorption is more complicated. Even if a gas absorbs radiation efficiently at a certain wavelength, this may not affect its GWP much if the atmosphere already absorbs most radiation at that wavelength. A gas has the most effect if it absorbs in a "window" of wavelengths where the atmosphere is fairly transparent. The dependence of GWP as a function of wavelength has been found empirically and published as a graph.

Because the GWP of a greenhouse gas depends directly on its infrared spectrum, the use of infrared spectroscopy to study greenhouse gases is centrally important in the effort to understand the impact of human activities on global climate change.

Calculating the Global-warming Potential

Just as radiative forcing provides a simplified means of comparing the various factors that are believed to influence the climate system to one another, global-warming potentials (GWPs) are one type of simplified index based upon radiative properties that can be used to estimate the potential future impacts of emissions of different gases upon the climate system in a relative sense. GWP is based on a number of factors, including the radiative efficiency (infrared-absorbing ability) of each gas relative to that of carbon dioxide, as well as the decay rate of each gas (the amount removed from the atmosphere over a given number of years) relative to that of carbon dioxide. The radiative forcing capacity (RF) is the amount of energy per unit area, per unit time, absorbed by the greenhouse gas, that would otherwise be lost to space. It can be expressed by the formula:

$$RF = \sum_{n=1}^{100} Abs_i * F_i / (pathlength * density)$$

where the subscript i represents an interval of 10 inverse centimeters. Abs_i represents the integrated infrared absorbance of the sample in that interval, and F_i represents the RF for that interval.

The Intergovernmental Panel on Climate Change (IPCC) provides the generally accepted values for GWP, which changed slightly between 1996 and 2001. An exact definition of how GWP is calculated is to be found in the IPCC's 2001 Third Assessment Report. The GWP is defined as the ratio of the time-integrated radiative forcing from the instantaneous release of 1 kg of a trace substance relative to that of 1 kg of a reference gas:

$$GWP(x) = \frac{\int_0^{TH} a_x \cdot [x(t)]\,dt}{\int_0^{TH} a_r \cdot [r(t)]\,dt}$$

where TH is the time horizon over which the calculation is considered; a_x is the radiative efficiency due to a unit increase in atmospheric abundance of the substance (i.e., $Wm^{-2}\ kg^{-1}$) and [x(t)] is the time-dependent decay in abundance of the substance following an instantaneous release of it at time t=0. The denominator contains the corresponding quantities for the reference gas (i.e. CO_2). The radiative efficiencies a_x and a_r are not necessarily constant over time. While the absorption of infrared radiation by many greenhouse gases varies linearly with their abundance, a few important ones display non-linear behaviour for current and likely future abundances (e.g., CO_2, CH_4, and N_2O). For those gases, the relative radiative forcing will depend upon abundance and hence upon the future scenario adopted.

Since all GWP calculations are a comparison to CO_2 which is non-linear, all GWP values are affected. Assuming otherwise as is done above will lead to lower GWPs for other gases than a more detailed approach would.

Use in Kyoto Protocol

Under the Kyoto Protocol, the Conference of the Parties decided (decision 2/CP.3) that the values of GWP calculated for the IPCC Second Assessment Report are to be used for converting the various greenhouse gas emissions into comparable CO_2 equivalents when computing overall sources and sinks.

Importance of Time Horizon

Note that a substance's GWP depends on the timespan over which the potential is calculated. A gas which is quickly removed from the atmosphere may initially have a large effect but for longer time periods as it has been removed becomes less important. Thus methane has a potential of 25 over 100 years but 72 over 20 years; conversely sulfur hexafluoride has a GWP of 22,800 over 100 years but 16,300 over 20 years (IPCC TAR). The GWP value depends on how the gas concentration decays over time in the atmosphere. This is often not precisely known and hence the values should not be considered exact. For this reason when quoting a GWP it is important to give a reference to the calculation.

The GWP for a mixture of gases can not be determined from the GWP of the constituent gases by any form of simple linear addition.

Commonly, a time horizon of 100 years is used by regulators (e.g., the California Air Resources Board).

Values

Carbon dioxide has a GWP of exactly 1 (since it is the baseline unit to which all other greenhouse gases are compared).

Although water vapour has a significant influence with regard to absorbing infrared radiation (which is the green house effect; see greenhouse gas), its GWP is not calculated. Its concentration in the atmosphere mainly depends on air temperature. There is no possibility to directly influence atmospheric water vapour concentration.

Hemispherical Photography

Hemispherical photography, also known as fisheye or canopy photography, is a technique to estimate solar radiationand characterize plant canopy geometry using photographs taken looking upward through an extreme wide-angle lens (Rich 1990). Typically, the viewing angle approaches or equals 180-degrees, such that all sky directions are simultaneously visible. The resulting photographs record the geometry of visible sky, or conversely the geometry of sky obstruction by plant canopies or other near-ground features. This geometry can be measured precisely and used to calculate solar radiation transmitted through (or intercepted by) plant canopies, as well as to estimate aspects of canopy structure such as leaf area index. Detailed treatments of field and analytical methodology have been provided by Paul Rich (1989, 1990) and Robert Pearcy (1989).

History of Hemispherical Photography

The hemispherical lens (also known as a fisheye or whole-sky lens) was originally designed by Robin Hill (1924) to view the entire sky for meteorological studies of cloud formation. Foresters and ecologists conceived of using photographic techniques to study the light environment in forests by examining the canopy geometry. In particular, Evans and Coombe (1959) estimated sunlight penetration through forest canopy openings by overlaying diagrams of the sun track on hemispherical photographs.

Later, Margaret Anderson (1964, 1971) provided a thorough theoretical treatment for calculating the transmission of direct and diffuse components of solar radiation through canopy openings using hemispherical photographs. At that time hemispherical photograph analysis required tedious manual scoring of overlays of sky quadrants and the track of the sun. With the advent of personal computers, researchers developed digital techniques for rapid analysis of hemispherical photographs (Chazdon and Field 1987, Rich 1988, 1989, 1990, Becker et al. 1989). In recent years, researchers have started using digital cameras in favour of film cameras, and algorithms are being developed for automated image classification and analysis. Various commercial software programmes have become available for hemispherical photograph analysis, and the technique has been applied for diverse uses in ecology, meteorology, forestry, and agriculture.

Applications of Hemispherical Photography

Hemispherical photography has been used successfully in a broad range of applications involving microsite characterization and estimation of the solar radiation near the ground and beneath plant canopies. For example, hemispherical photography has been used to characterize winter roosting sites for monarch butterflies (Weiss et al. 1991), effects of forest edges (Galo et al. 1991), influence of forest treefall gaps on tree regeneration (Rich et al. 1993), spatial and temporal variability of light in tropical rainforest understory (Clark et al. 1996), impacts of hurricanes on forest ecology (Bellingham et al. 1996), leaf area index for validation of remote sensing (Chen et al. 1997), canopy architecture of boreal forests (Fournier et al. 1997), light environment in old growth temperate rain forests (Weiss 2000), and management of vineyard trellises to make better wine (Weiss et al. 2003).

Theory of Hemispherical Photography

Solar Radiation Calculations

Direct and diffuse components of solar radiation are calculated separately. Direct radiation is calculated as the sum of all direct (solar beam) radiation originating from visible (non-obscured) sky directions along the path of the sun. Similarly, diffuse solar radiation is calculated as the sum of all diffuse radiation (scattered from the atmosphere) originating from any visible (non-obscured) sky directions. The sum of direct and diffuse components gives global radiation.

These calculations require theoretical or empirical distributions of direct and diffuse radiation in the open, without canopy or other sky obstruction. Usually calculations are performed for either photosynthetically active radiation (400-700 nanometers) or insolation integrated over all wavelengths, measured in kilowatt-hours per square metre (kW h/m^2).

The fundamental assumption is that most solar radiation originates from visible (unobscured) sky directions, a strong first order effect, and that reflected radiation from the canopy or other near-ground features (non-visible or obscured sky directions) is negligible, a small second order effect. Another assumption is that the geometry of visible (non-obscured) sky does not change over the period for which calculations are performed.

Canopy Calculations

Canopy indices, such as leaf area index, are based on calculation of gap fraction, the proportion of visible (non-obscured) sky as a function of sky direction. Leaf area index is typically calculated as the leaf area per unit ground area that would produce the observed gap fraction distribution, given an assumption of random leaf angle distribution, or a known leaf angle distribution and degree of clumping. For further explanation see leaf area index.

Indices

Direct Site Factor (DSF) is the proportion of direct solar radiation at a given location relative to that in the open, either integrated over time or resolved according to intervals of time of day and/or season.

Indirect Site Factor (ISF) is the proportion of diffuse solar radiation at a given location relative to that in the open, either integrated over time for all sky directions or resolved by sky sector direction.

Global Site Factor (GSF) is the proportion of global solar radiation at a given location relative to that in the open, calculated as the sum of DSF and ISF weighted by the relative contribution of direct versus diffuse components. Indices may be uncorrected or corrected for angle of incidence relative to a flat intercepting surface. Uncorrected values weight solar radiation originating from all directions equally. Corrected values weight solar radiation by the cosine of the angle of incidence, accounting for actual interception from directions normal to the intercepting surface.

Leaf Area Index is the total leaf surface area per unit ground area.

Methodology

Hemispherical photography entails five steps: photograph acquisition, digitization, registration, classification, and calculation. Registration, classification, and calculation are accomplished using dedicated hemispherical photography analysis software.

Photograph Acquisition

Upward-looking hemispherical photographs are typically acquired under uniform sky lighting, early or late in the day or under overcast conditions. Known orientation (zenith and azimuth) is essential for proper registration with the analysis hemispherical coordinate system. Even lighting is essential for accurate image classification. A self-levelling mount (gimbals) can facilitate acquisition by ensuring that the camera is oriented to point straight up towards the zenith. The camera is typically oriented such that north (absolute or magnetic) is oriented towards the top of the photograph.

The lens used in hemispherical photography is generally a circular fisheye lens, such as the Nikkor 8mm fisheye lens. Full-frame fisheye lenses are *not* suitable for hemispherical photography, as they only capture a full 180° across the diagonal, and do not provide a complete hemispherical view.

In the early years of the technique, most hemispherical photographs were acquired with 35 mm cameras (e.g., Nikon FM2 with a Nikkor 8mm fisheye lens) using high contrast, high ASA black-and-white film. Later, use of colour film or slides became common. Recently most photographs are acquired using digital cameras (e.g., Kodak DCS Pro 14nx with a Nikkor 8mm fisheye lens).

When images are acquired from locations with large differences in openness (for example, closed canopy locations and canopy gaps) it is essential to control camera exposure. If the camera is allowed to automatically adjust exposure (which is controlled by aperture and shutter speed), the result is that small openings in closed conditions will be bright, whereas openings of the same size in open conditions will be darker (for example, canopy areas around a gap). This means that during image analysis the same sized holes will be interpreted as "sky" in a closed-canopy image and "canopy" in the open-canopy image. Without controlling exposure, the real differences between closed- and open-canopy conditions will be under-estimated.

Digitization

Photographs are digitized and saved in standard image formats. For film cameras this step requires a negative or slide scanner or a video digitizer. For digital cameras this step occurs as photographs are acquired.

Registration

Photograph registration involves aligning the photographs with the hemispherical coordinate system used for analysis, in terms of translation (centering), size (coincidence of photograph edges and horizon in coordinate system), and rotation (azimuthal alignment with respect to compass directions).

Classification

Photograph classification involves determining which image pixels represent visible (non-obscured) versus non-visible (obscured) sky directions. Typically this has been accomplished using interactive thresholding, whereby an appropriate threshold is selected to best match a binary classification with observed sky visibility, with pixel intensity values above the threshold classified as visible and pixel intensity values below the threshold classified as non-visible. Recently advances have been made in developing automatic threshold algorithms, however more work is still needed before these are fully reliable.

Calculation

Hemispherical photograph calculation uses algorithms that compute gap fraction as function of sky direction, and compute desired canopy geometry and/or solar radiation indices. For solar radiation, rapid calculation is often accomplished using pre-calculated lookup tables of theoretical or empirical solar radiation values resolved by sky sector or position in the sunpath.

Radiative Forcing

In climate science, radiative forcing is generally defined as the change in net irradiance between different layers of the atmosphere. Typically, radiative forcing is quantified at thetropopause in units of watts per square metre. A positive forcing (more incoming energy) tends to warm the system, while a negative forcing (more outgoing energy) tends to cool it. Sources of radiative forcing include changes in insolation (incident solar radiation) and in concentrations of radiatively active gases and aerosols.

Radiation Balance

The vast majority of the energy which affects Earth's weather comes from the Sun. The planet and its atmosphere absorb and reflect some of the energy, while long-wave energy is radiated back into space. The balance between absorbed and radiated energy determines the average temperature. The planet is warmer than it would be in the absence of the atmosphere: see greenhouse effect.

The radiation balance can be altered by factors such as intensity of solar energy, reflection by clouds or gases, absorption by various gases or surfaces, emission of heat by various materials, and other factors related to climate change. Any such alteration is a radiative forcing, and causes a new balance to be reached. In the real world this happens continuously as sunlight hits the surface, clouds and aerosols form, the concentrations of atmospheric gases vary, and seasons alter the ground cover.

IPCC usage

The term "radiative forcing" has been used in the IPCC Assessments with a specific technical meaning, to denote an externally imposed perturbation in the radiative energy budget of Earth's climate system, which may lead to changes in climate parameters. The exact definition used is:

The radiative forcing of the surface-troposphere system due to the perturbation in or the introduction of an agent (say, a change in greenhouse gas concentrations) is the change in net (down minus up) irradiance (solar plus long-wave; in Wm^{-2}) at the tropopause AFTER allowing for stratospheric temperatures to readjust to radiative equilibrium, but with surface and tropospheric temperatures and state held fixed at the unperturbed values.

In a subsequent report, the IPCC defines it as:

> *"Radiative forcing is a measure of the influence a factor has in altering the balance of incoming and outgoing energy in the Earth-atmosphere system and is an index of the importance of the factor as a potential climate change mechanism. In this report radiative forcing values are for changes relative to preindustrial conditions defined at 1750 and are expressed in Watts per square metre (W/m^2)."*

In simple terms, radiative forcing is "...the rate of energy change per unit area of the globe as measured at the top of the atmosphere." In the context of climate change, the term "forcing" is restricted to changes in the radiation balance of the surface-troposphere system imposed by external factors, with no changes in stratospheric dynamics, no surface and tropospheric feedbacks in operation (*i.e.*, no secondary effects induced because of changes in tropospheric motions or its thermodynamic state), and no dynamically induced changes in the amount and distribution of atmospheric water (vapour, liquid, and solid forms).

Climate Sensitivity

Climate sensitivity is a measure of how responsive the temperature of the climate system is to a change in the radiative forcing.

Although climate sensitivity is usually used in the context of radiative forcing by carbon dioxide, it is thought of as a general property of the climate system: the change in surface air temperature (ΔT_s) following a unit change in radiative forcing (RF), and thus is expressed in units of °C/(W/m^2). For this to be useful, the measure must be independent of the nature of the forcing (e.g. from greenhouse gases or solar variation); to first order this is indeed found to be so. The climate sensitivity specifically due to CO_2 is often expressed as the temperature change in °C associated with a doubling of the concentration of carbon dioxide in Earth's atmosphere. For a coupled atmosphere-ocean global climate model the climate sensitivity is an emergent property:

it is not a model parameter, but rather a result of a combination of model physics and parameters. By contrast, simpler energy-balance models may have climate sensitivity as an explicit parameter.

$$\Delta T_s = \lambda \cdot RF$$

The terms represented in the equation relate radiative forcing of any cause to linear changes in global surface temperature change. It is also possible to estimate climate sensitivity from observations; however, this is difficult due to uncertainties in the forcing and temperature histories.

Equilibrium and Transient Climate Sensitivity

The equilibrium climate sensitivity (ECS) refers to the equilibrium change in global mean near-surface air temperature that would result from a sustained doubling of the atmospheric (equivalent) carbon dioxide concentration (ΔT_{x2}). This value is estimated, by the IPCC Fourth Assessment Report (*AR4*) as *likely to be in the range 2 to 4.5 °C with a best estimate of about 3 °C, and is very unlikely to be less than 1.5 °C. Values substantially higher than 4.5 °C cannot be excluded, but agreement of models with observations is not as good for those values*. This is a slight change from the IPCC Third Assessment Report (*TAR*), which said it was "likely to be in the range of 1.5 to 4.5 °C–. Some subsequent work continues to support a most likely value around 3 °C, but one paper published in 2011 by Andreas Schmittner, et. al. suggests a lower median value of 2.3 K.

A model estimate of equilibrium sensitivity thus requires a very long model integration. A measure requiring shorter integrations is the transient climate response (TCR) which is defined as the average temperature response over a twenty year period centered at CO_2 doubling in a transient simulation with CO_2 increasing at 1% per year. The transient response is lower than the equilibrium sensitivity, due to the "inertia" of ocean heat uptake. Fully equilibrating ocean temperatures requires integrations of thousands of model years.

Over the 50–100 year timescale, the climate response to forcing is likely to follow the TCR; for considerations of climate stabalisation, the ECS is more useful.

An estimate of the equilibrium climate sensitivity may be made from combining the effective climate sensitivity with the known properties of the ocean reservoirs and the surface heat fluxes; this is the effective climate sensitivity. This "may vary with forcing history and climate state".

A less commonly used concept, the Earth system sensitivity (ESS), can be defined which includes the effects of slower feedbacks, such as the albedo change from melting the large ice sheets that covered much of the northern hemisphere during the last glacial maximum. These extra feedbacks make the ESS larger than the ECS — possibly twice as large — but also mean that it may well not apply to current conditions.

Sensitivity to Carbon Dioxide Forcing

Radiative Forcing Due to Doubled CO_2

CO_2 climate sensitivity has a component directly due to radiative forcing by CO_2, and a further contribution arising from feedbacks, positive and negative. "Without any feedbacks, a doubling of CO_2 (which amounts to a forcing of 3.7 W/m^2) would result in 1 °C global warming, which is easy to calculate and is undisputed. The remaining uncertainty is due entirely to feedbacks in the system, namely, the water vapor feedback, the ice-albedo feedback, the cloud feedback, and the lapse rate feedback"; addition of these feedbacks leads to a value of the sensitivity to CO_2 doubling of approximately 3 °C ± 1.5 °C, which corresponds to a value of λ of 0.8 K/(W/m^2).

In the earlier 1979 NAS report (p. 7), the radiative forcing due to doubled CO_2 is estimated to be 4 W/m^2, as calculated (for example) in Ramanathan et al. (1979). In 2001 the IPCC adopted the revised value of 3.7 W/m^2, the difference attributed to a "stratospheric temperature adjustment". More recently an intercomparison of radiative transfer codes (Collins et al., 2006) showed discrepancies among climate models and between climate models and more exact radiation codes in the forcing attributed to doubled CO_2 even in cloud-free sky; presumably the differences would be even greater if forcing were evaluated in the presence of clouds because of differences in the treatment of clouds in different models. Undoubtedly the difference in forcing attributed to doubled CO_2 in different climate models contributes to differences in apparent sensitivities of the models, although this effect is thought to be small relative to the intrinsic differences in sensitivities of the models themselves (Webb et al., 2006).

Three Degrees as the Consensus Estimate

A committee on anthropogenic global warming convened in 1979 by the National Academy of Sciences and chaired by Jule Charney estimated climate sensitivity to be 3 °C, plus or minus 1.5 °C. Only two sets of models were available; one, due to Syukuro Manabe, exhibited a climate sensitivity of 2 °C, the other, due to James E.

Hansen, exhibited a climate sensitivity of 4 °C. "According to Manabe, Charney chose 0.5 °C as a not-unreasonable margin of error, subtracted it from Manabe's number, and added it to Hansen's. Thus was born the 1.5 °C-to-4.5 °C range of likely climate sensitivity that has appeared in every greenhouse assessment since..." Chapter 4 of the "Charney report" compares the predictions of the models: "We conclude that the predictions ... are basically consistent and mutually supporting. The differences in model results are relatively small and may be accounted for by differences in model characteristics and simplifying assumptions."

The 1990 IPCC *First Assessment Report* estimated that equilibrium climate sensitivity to CO_2 doubling lay between 1.5 and 4.5 °C, with a "best guess in the light of current knowledge" of 2.5 °C . This used models with strongly simplified representations of the ocean dynamics. The IPCC supplementary report, 1992 which used full ocean GCMs nonetheless saw "no compelling reason to warrant changing" from this estimate and the IPCC Second Assessment Report found that "No strong reasons have emerged to change" these estimates, with much of the uncertainty attributed to cloud processes. As noted above, the IPCC TAR retained the 1.5 to 4.5 °C, and the AR4 tightened it slightly to 2 to 4.5 °C with a best estimate of about 3 °C.

In 2008 climatologist Stefan Rahmstorf wrote, regarding the Charney report's original range of uncertainty: "At that time, this range was on very shaky ground. Since then, many vastly improved models have been developed by a number of climate research centres around the world. Current state-of-the-art climate models span a range of 2.6–4.1 °C, most clustering around 3 °C."

Calculations of CO_2 Sensitivity from Observational Data

Sample Calculation Using Industrial-age Data

Rahmstorf (2008) provides an informal example of how climate sensitivity might be estimated empirically, from which the following is modified. Denote the sensitivity, i.e. the equilibrium increase in global mean temperature including the effects of feedbacks due to a sustained forcing by doubled CO_2 (taken as 3.7 W/m^2), as x °C. If Earth were to experience an equilibrium temperature change of ΔT (°C) due to a sustained forcing of ΔF (W/m^2), then one might say that $x/(\Delta T) = (3.7 \text{ W/m}^2)/(\Delta F)$, i.e. that $x = \Delta T * (3.7 \text{ W/m}^2)/\Delta F$. The global temperature increase since the beginning of the industrial period (taken as 1750) is about 0.8 °C, and the radiative forcing due to CO_2 and other long-lived greenhouse gases (mainly methane, nitrous oxide, and chlorofluorocarbons) emitted since that time is about 2.6 W/m^2. Neglecting other forcings and considering the temperature increase to be an equilibrium increase would lead to a sensitivity of about 1.1 °C.

However, ΔF also contains contributions due to solar activity (+0.3 W/m^2), aerosols (-1 W/m^2), ozone (0.3 W/m^2) and other lesser influences, bringing the total forcing over the industrial period to 1.6 W/m^2 according to best estimate of the IPCC AR4, albeit with substantial uncertainty. Additionally the fact that the climate system is not at equilibrium must be accounted for; this is done by subtracting the planetary heat uptake rate H from the forcing; i.e., $x = \Delta T * (3.7\ W/m^2)/(\Delta F\text{-}H)$. Taking planetary heat uptake rate as the rate of ocean heat uptake, estimated by the IPCC AR4 as 0.2 W/m^2, yields a value for x of 2.1 °C. (All numbers are approximate and quite uncertain.)

Sample Calculation Using Ice-age Data

In 2008, Farley wrote: "... examine the change in temperature and solar forcing between glaciation (ice age) and interglacial (no ice age) periods. The change in temperature, revealed inice core samples, is 5 °C, while the change in solar forcing is 7.1 W/m^2. The computed climate sensitivity is therefore 5/7.1 = 0.7 K(W/m^2)$^{-1}$. We can use this empirically derived climate sensitivity to predict the temperature rise from a forcing of 4 W/m^2, arising from a doubling of the atmospheric CO_2 from pre-industrial levels. The result is a predicted temperature increase of 3 °C."

Based on analysis of uncertainties in total forcing, in Antarctic cooling, and in the ratio of global to Antarctic cooling of the last glacial maximum relative to the present, Ganopolski and Schneider von Deimling (2008) infer a range of 1.3 to 6.8 °C for climate sensitivity determined by this approach.

A lower figure was calculated in a 2011 *Science* paper by Schmittner *et al.*, who combined temperature reconstructions of the Last Glacial Maximum with climate model simulations to suggest a rate of global warming from doubling of atmospheric carbon dioxide of a median of 2.3 °C and uncertainty 1.7–2.6 °C (66% probability range), less than the earlier estimates of 2 to 4.5 °C as the 66% probability range). Schmittner et al. said their "results imply less probability of extreme climatic change than previously thought." Their work suggests that climate sensitivities >6 °C "cannot be reconciled with paleoclimatic and geologic evidence, and hence should be assigned near-zero probability."

Other Experimental Estimates

Idso (1998) calculated based on eight natural experiments a λ of 0.1 °C/(Wm^{-2}) resulting in a climate sensitivity of only 0.4 °C for a doubling of the concentration of CO_2 in the atmosphere.

Andronova and Schlesinger (2001) found that the climate sensitivity could lie between 1 and 10 °C, with a 54 percent likelihood that it lies outside the IPCC range. The exact range depends on which factors are most important during the instrumental period: "At present, the most likely scenario is one that includes anthropogenic sulfate aerosol forcing but not solar variation. Although the value of the climate sensitivity in that case is most uncertain, there is a 70 percent chance that it exceeds the maximum IPCC value. This is not good news," said Schlesinger. Forest, *et al.* (2002) using patterns of change and the MIT EMIC estimated a 95% confidence interval of 1.4–7.7 °C for the climate sensitivity, and a 30% probability that sensitivity was outside the 1.5 to 4.5 °C range.

Gregory, *et al.* (2002) estimated a lower bound of 1.6 °C by estimating the change in Earth's radiation budget and comparing it to the global warming observed over the 20th century.

Shaviv (2005) carried out a similar analysis for 6 different time scales, ranging from the 11-yr solar cycle to the climate variations over geological time scales. He found a typical sensitivity of 0.54±0.12 K/(W m^{-2}) or 2.1 °C (ranging between 1.6 °C and 2.5 °C at 99% confidence) if there is no cosmic-ray climate connection, or a typical sensitivity of 0.35±0.09 K/(W m^{-2}) or 1.3 °C (between 1.0 °C and 1.7 °C at 99% confidence), if the cosmic-ray climate link is real. (Note Shaviv quotes a radiative forcing equivalent of 3.8Wm^{-2}. [ΔT_{x2}=3.8 Wm$^{-2}\lambda$].) Frame, *et al.* (2005) noted that the range of the confidence limits is dependent on the nature of the prior assumptions made.

Annan and Hargreaves (2006) presented an estimate that resulted from combining prior estimates based on analyses of paleoclimate, responses to volcanic eruptions, and the temperature change in response to forcings over the twentieth century. They also introduced a triad notation (L, C, H) to convey the probability distribution function (pdf) of the sensitivity, where the central value C indicates the maximum likelihood estimate in degrees Celsius and the outer values L and H represent the limits of the 95% confidence interval for a pdf, or 95% of the area under the curve for a likelihood function. In this notation their estimate of sensitivity was (1.7, 2.9, 4.9)°C.

Forster and Gregory (2006) presented a new independent estimate based on the slope of a plot of calculated greenhouse gas forcing minus top-of-atmosphere energy imbalance, as measured by satellite borne radiometers, versus global mean surface temperature. In the triad notation of Annan and Hargreaves their estimate of sensitivity was (1.0, 1.6, 4.1)°C.

Royer, *et al.* (2007) determined climate sensitivity within a major part of the Phanerozoic. The range of values—1.5 °C minimum, 2.8 °C best estimate, and 6.2 °C maximum—is, given various uncertainties, consistent with sensitivities of current climate models and with other determinations.

Radiative Forcing Functions

A number of different inputs can give rise to radiative forcing. In addition to the downwelling radiation due to the greenhouse effect, the IPCC First Scientific Assessment Report listed solar radiation variability due to orbital changes, variability due to changes in solar irradiance, direct aerosol effects (*e.g.*, changes in albedo due to cloud cover), indirect aerosol effects, and surface characteristics.

Sensitivity to Solar Forcing

Solar luminosity is about 0.9 W/m^2 brighter during solar maximum than during solar minimum. Analysis by Camp and Tung shows that this correlates with a variation of ±0.1°C in measured average global temperature between the peak and minimum of the 11-year solar cycle. From this data (incorporating the Earth's albedo and the fact that the solar absorption cross-section is 1/4 of the surface area of the Earth), Tung, Zhou and Camp (2008) derive a transient sensitivity value of 0.69 to 0.97 °C/(W/m^2). This would correspond to a transient climate sensitivity to carbon dioxide doubling of 2.5 to 3.6 K, similar to the range of the current scientific consensus. However, they note that this is the transient response to a forcing with an 11 year cycle; due to lag effects, they estimate the equilibrium response to forcing would be about 1.5 times higher.

Example calculations:

Solar Forcing

Radiative forcing (measured in Watts per square metre) can be estimated in different ways for different components. For the case of a change in solar irradiance (*i.e.*, "solar forcing"), the radiative forcing is simply the change in the average amount of solar energy absorbed per square metre of the Earth's area. Since the cross-sectional area of the Earth exposed to the sun (πr^2) is equal to 1/4 of the surface area of the Earth ($4\pi r^2$), the solar input per unit area is one quarter the change in solar intensity.

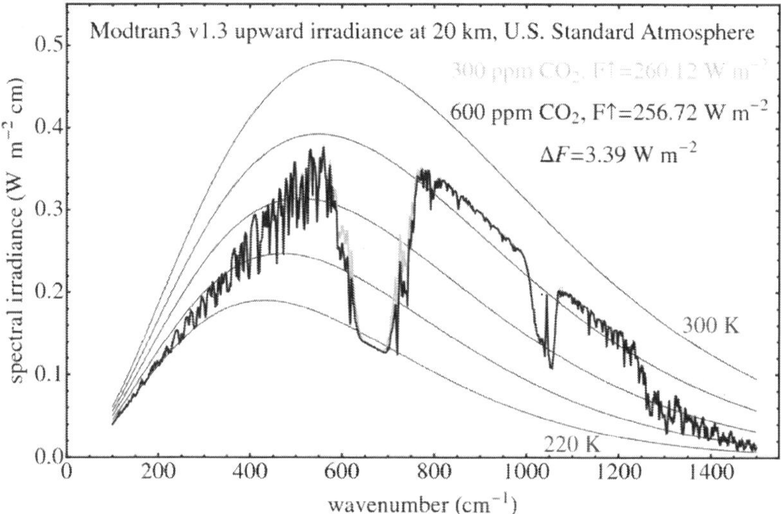

Figure: *Radiative forcing for doubling CO_2, as calculated by radiative transfer code Modtran. Red lines are Planck curves.*

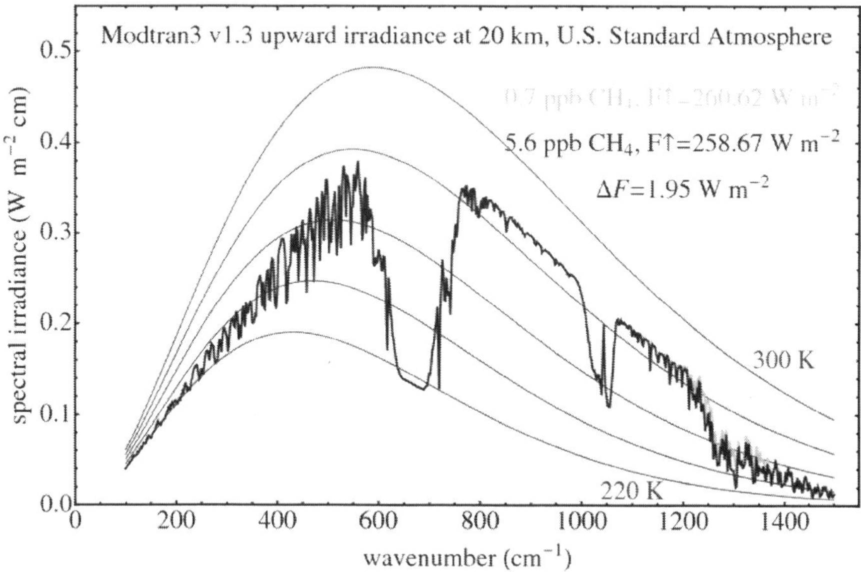

Figure: *Radiative forcing for eight times increase of CH_4, as calculated by radiative transfer code Modtran.*

This must be multiplied by the fraction of incident sunlight that is absorbed, F=(1-R), where R is the reflectivity, or albedo, of the Earth, equal to approximately 0.7. Thus, the solar forcing is the change in the solar intensity divided by 4 and multiplied by 0.7.

Likewise, a change in albedo will produce a solar forcing equal to the change in albedo divided by 4 multiplied by the solar constant.

Forcing Due to Atmospheric Gas

For a greenhouse gas, such as carbon dioxide, radiative transfer codes that examine each spectral line for atmospheric conditions can be used to calculate the change ΔF as a function of changing concentration. These calculations can often be simplified into an algebraic formulation that is specific to that gas.

For instance, the simplified first-order approximation expression for carbon dioxide is:

$$\Delta F = 5.35 \times \ln \frac{C}{C_0} \mathrm{W\,m^{-2}}$$

where C is the CO_2 concentration in parts per million by volume and C_0 is the reference concentration. The relationship between carbon dioxide and radiative forcing is logarithmic, and thus increased concentrations have a progressively smaller warming effect.

The same logarithmic formula applies for other greenhouse gases such as methane, N_2O or CFCs, with coefficients that can be found*e.g.* in the IPCC reports.

Related Measures

Radiative forcing is intended as a useful way to compare different causes of perturbations in a climate system. Other possible tools can be constructed for the same purpose: for example Shine *et al.* say "...recent experiments indicate that for changes in absorbing aerosols and ozone, the predictive ability of radiative forcing is much worse... we propose an alternative, the 'adjusted troposphere and stratosphere forcing'.

We present GCM calculations showing that it is a significantly more reliable predictor of this GCM's surface temperature change than radiative forcing. It is a candidate to supplement radiative forcing as a metric for comparing different mechanisms...". In this quote, GCM stands for "global circulation model", and the word "predictive" does not refer to the ability of GCMs to forecast climate change. Instead, it refers to the ability of the alternative tool proposed by the authors to help explain the system response.

Abrupt Climate Change

An abrupt climate change occurs when the climate system is forced to transition to a new state at a rate that is determined by the climate system itself, and which is more rapid than the rate of change of the external forcing. Past events include the end of the Carboniferous Rainforest Collapse, Younger Dryas, Dansgaard-Oeschger events, and possibly also the Paleocene-Eocene thermal maximum. The term is also used within the context of global warming to describe sudden climate change that is detectable over the time-scale of a human lifetime. One proposed reason for the observed abrupt climate change is that feedback loops within the climate system both enhance small perturbations and cause a variety of stable states.

Timescales of events described as 'abrupt' may vary dramatically. Changes recorded in the climate of Greenland at the end of the Younger Dryas, as measured by ice-cores, imply a sudden warming of +10°C within a timescale of a few years. Other abrupt changes are the +4 °C on Greenland 11,270 years ago or the abrupt +6 °C warming 22 000 years ago on Antarctica. By contrast, the Paleocene-Eocene Thermal Maximum may have initiated anywhere between a few decades and several thousand years.

Definitions

According to the Committee on Abrupt Climate Change of the National Research Council.

There are essentially two definitions of abrupt climate change:

- In terms of physics, it is a *transition of the climate system into a different mode on a time scale that is faster than the responsible forcing.*
- In terms of impacts, *"an abrupt change is one that takes place so rapidly and unexpectedly that human or natural systems have difficulty adapting to it".*

These definitions are complementary: the former gives some insight into how abrupt climate change comes about ; the latter explains why there is so much research devoted to it, why it inspires catastrophe movies, and may even be the reason why you are reading this page.

Current Situation

The IPCC states that global warming "could lead to some effects that are abrupt or irreversible". In an article in Science, Alley et al. said "it is conceivable that human forcing of climate change is increasing the probability of large, abrupt events. Were such an event to recur, the economic and ecological impacts could be large and potentially serious."

Regional Changes

Lenton et al. investigated tipping elements in the climate system. These were regional effects of global warming, some of which had abrupt onset and may therefore be regarded as abrupt climate change. They found that "Our synthesis of present knowledge suggests that a variety of tipping elements could reach their critical point within this century under anthropogenic climate change."

Ocean Effects

Global oceans have established patterns of currents. Several potential disruptions to this system of currents have been identified as a result of global warming:

- Increasing frequency of El Niño events.
- Potential disruption to the thermohaline circulation, such as that which may have occurred during the Younger Dryas event.
- Changes to the North Atlantic oscillation

Climate Feedback Effects

One source of abrupt climate change effects is a feedback process, in which a warming event causes a change which leads to further warming. This can also apply to cooling. Example of such feedback processes are:

- Ice-albedo feedback, where the advance or retreat of ice cover alters the 'whiteness' of the earth, and its ability to absorb the sun's energy.
- The dying and burning of forests, as a result of global warming.

Past Events

Several periods of abrupt climate change have been identified in the paleoclimatic record. Notable examples include:

- About 25 climate shifts, called Dansgaard-Oeschger cycles, which have been identified in the ice core record during the glacial period over the past 100,000 years. The most recent of these events was the Younger Dryas which began 12,900 years ago and moved back into a warm-and-wet climate regime about 11,600 years ago.
- The Younger Dryas event, notably its sudden end. It has been suggested that: "The extreme rapidity of these changes in a

variable that directly represents regional climate implies that the events at the end of the last glaciation may have been responses to some kind of threshold or trigger in the North Atlantic climate system." A model for this event based on disruption to the thermohaline circulation has been supported by other studies.

- The Paleocene-Eocene Thermal Maximum, timed at 55 million years ago, which may have been caused by the clathrate gun effect, although potential alternative mechanisms have been identified. This was associated with rapid ocean acidification
- The Permian-Triassic Extinction Event, also known as the great dying, in which up to 95% of all species became extinct, has been hypothesized to be related to a rapid change in global climate. Life on land took 30M years to recover.
- The Carboniferous Rainforest Collapse occurred 300 million years ago, at which time tropical rainforests were devastated by climate change. The cooler, drier climate had a severe effect on the biodiversity of amphibians, the primary form of vertebrate life on land.

There are also abrupt climate changes associated with the catastrophic draining of glacial lakes. One example of this is the 8.2 kiloyear event, which associated with the draining of Glacial Lake Agassiz. Another example is the Antarctic Cold Reversal, c. 14,500 years before present (BP), which is believed to have been caused by a meltwater pulse from the Antarctic ice sheet. These rapid meltwater release events have been hypothesized as a cause for Dansgaard-Oeschger cycles,

Abrupt Climate Shifts Since 1976

Had the 1997 El Niño lasted twice as long, the rain forests of the Amazon basin and Southeast Asia could have quickly added much additional carbon dioxide to the air from burning and rotting, with heat waves and extreme weather quickly felt around the world (The "Burn Locally, Crash Globally" scenario.)

Most abrupt climate shifts, however, are likely due to sudden circulation shifts, analogous to a flood cutting a new river channel. The best-known examples are the several dozen shutdowns of the North Atlantic Ocean's Meridional Overturning Circulation during the last ice age, affecting climate worldwide. But there have been a series of less dramatic abrupt climate shifts since 1976, along with some near misses.

- The circulation shift in the western Pacific in the winter of 1976-1977 proved to have much wider impacts.
- Since 1950, El Niòos had been weak and short, but La Niòas were often big and long, This pattern reversed after 1977.
- Land temperatures had remained relatively trendless from 1950 to 1976, despite the CO_2 rising from 310 to 332 ppm as fossil fuel emissions tripled. Then in 1977 there was a marked shift in observed global mean surface temperature to a rising fever of about 2°C/century.
- The expansion of the tropics from overheating is usually thought to be gradual, but the percentage of the land surface in the two most extreme classifications of drought suddenly doubled in 1982 and stayed there until 1997 when it jumped to triple (after six years, it stepped down to double). While their inceptions correlate with the particularly large El Niòos of 1982 and 1997, the global drought steps far outlast the 13-month durations of those El Niòos.
- There were near-misses for Burn Locally, Crash Globally in Amazonia in 1998, 2005, and 2007, each with higher flammability than its predecessor.
- There have also been two occasions when the Atlantic's Meridional Overturning Circulation lost a crucial safety factor. The Greenland Sea flushing at 75 °N shut down in 1978, recovering over the next decade. Then the second-largest flushing site, the Labrador Sea, shut down in 1997 for ten years. While shutdowns overlapping in time have not been seen during the fifty years of observation, previous total shutdowns had severe worldwide climate consequences.

This makes abrupt climate shifts more like a heart attack than like a chronic disease whose course can be extrapolated. Like heart attacks, some abrupt climate shifts are minor, some are catastrophic—and one cannot predict which or when. The recent track record, however, is that there have been several sudden shifts and several near-misses in each decade since 1976.

Millions of Years Ago

The Permian–Triassic extinction event, labelled "P-Tr" here, is the most significant extinction event in this plot for marinegenera.

Abrupt climate change has likely been the cause of wide ranging and severe effects:

- Loss of biodiversity. Human impact has changed the pattern of diversification. Without human interference the biodiversity of this planet would continue to grow at an exponential rate.
- Rapid Ocean acidification, which can harm marine life (such as corals).
- Mass extinctions in the past, most notably the Permian-Triassic Extinction event (often referred to as the great dying) and the Carboniferous Rainforest Collapse, have been suggested as a consequence of abrupt climate change.

Alkenone

Alkenones are highly resistant organic compounds (ketones) produced by phytoplankton of the class Prymnesiophyceae. The exact function of the alkenones remains under debate. Coccolithophoroids, for instance *Emiliania huxleyi*, respond to changes in water temperature by altering the production of long-chain unsaturated alkenones in the structure of their cell. At higher temperatures, more of the di-unsaturated molecules are produced than tri-unsaturated [Prahl and Wakeham]. The molecules are resistant to diagenesis, and can be recovered from sediments up to 110 million years old.

The ambient water temperature in which the organisms dwelt can be estimated from ratio of their unsaturated alkenones (C_{37}-C_{39}) that are preserved in marine sediments. The modified Unsaturation Index of "di" versus "tri" unsaturated C_{37} alkenone is calculated according to the following relationship from [Prahl and Wakeham], which is modified after the original Unsaturation Index from [Brassell et al] that included the tetra-unsaturated alkenone:

$$U^{K2}_{37} = C_{37:2}/(C_{37:2} + C_{37:3})$$

The Unsaturation Index can then be used to estimate the water temperature according to the following experimental relationship [Prahl and Wakeham]:

$$T\,[°C] = (U^{K2}_{37} - 0.039)/0.034$$

Atlantic (period)

The Atlantic in palaeoclimatology was the warmest and moistest Blytt-Sernander period, pollen zone and chronozone of Holocenenorthern Europe. The climate was generally warmer than today. It was preceded by the Boreal, with a climate similar to today's, and was followed by the Sub-Boreal, a transition to the modern. Because it was the warmest period of the Holocene, the Atlantic is often referenced more directly as the Holocene climatic optimum, or just climatic optimum.

Subdividing the Atlantic

The Atlantic is equivalent to Pollen Zone VII. Sometimes a Pre-atlantic or early Atlantic is distinguished, on the basis of an early dividing cold snap. Other scientists place the Atlantic entirely after the cold snap, assigning the latter to the Boreal. The period is still in the process of definition.

Dating

Beginning of the Atlantic Period

It is a question of definition and the criteria: Beginning with the temperatures, as derivable from Greenland ice core data, it is possible to define an 'Early' or 'Pre-Atlantic' period at around 8,040 BC, where the ^{18}-isotope line remains above 33 ppm in the combined curve after Rasmussen et al. (2006), which then would end at the well-known 6,2 ka BC (8,2 ka calBP)-cold-event.

Or one single Atlantic period is defined, starting at that just mentioned cold-event. After a lake-level criterion, Kul'kova and others define the Atlantic as running from 8000 to 5000 (cal?) BP. Early Atlantic, or AT1, was a time of high lake levels, 8000–7000; in Middle Atlantic, AT2, lakes were at a lower level, 7000–6500; and in Late Atlantic I, 6500–6000, and II, 6000–5700, levels were on the rise. Each period has its distinctive ratios of species.

End of the Atlantic Period

According to the ice-core criterion it is extremely difficult to find a clear boundary, because the measurements still differ too much and alignments are still under construction. Many find a decline of temperature significant enough after 4800 BC. Another criterion comes from bio-stratigraphy: The elm-decline. However, this appears in different regions between 4300 and 3100 BC.

Description

The Atlantic was a time of rising temperature and marine transgression on the islands of Denmark and elsewhere. The sea rose to 3 m above its present level by the end of the period. The oysters found there required lower salinity. Tides of up to 1 m were present. Inland, lake levels in all north Europe were generally higher, with fluctuations.

The temperature rise had the effect of extending southern climates northward in a relatively short period. The treelines on northern mountains rose by 600 to 900 m (2000–3000 feet). Thermophilous ("heat-loving") species migrated northward. They did not replace the species that were there, but shifted the percentages in their favour. Across middle Europe, the boreal forests were replaced by climax or "old growth" deciduous ones, which, though providing a denser canopy, were more open at the base.

The dense canopy theory, however, has been questioned recently by F. Vera. Oak and hazel require more light than is allowed by the dense canopy. Vera hypothesizes that the lowlands were more open and that the low frequency of grass pollen was caused by the browsing of large herbivores, such as Bos primigenius and Equus ferus.

Flora

During the Atlantic period the deciduous temperate zone forests of south and central Europe extended northward to replace the boreal mixed forest, which found refugia on the mountain slopes. Mistletoe, Water Chestnut (*Trapa natans*) and Ivy (*Hedera helix*) were present in Denmark. Grass pollen decreased. Softwood forests were replaced by hardwood. *Quercus*, *Tilia*, both cordata and platyphyllos, beech, oak, hazel, linden, *Ulmus glabra*, alder, and ash replaced *Betula* and *Pinus*, spreading to the north from further south. The period is sometimes called "the alder-elm-lime period".

In northeast Europe, the Early Atlantic forest was but slightly affected by the rise in temperature. The forest had been pine with an underbrush of hazel, alder, birch, and willow. Only about 7% of the forest became broad-leaved deciduous, dropping to Boreal levels in the cooling of the Middle Atlantic. In the warmer Late Atlantic, the broad-leaved trees became 34% of the forest. Along the line of the Danube and the Rhine, extending northward in tributary drainage systems, a new factor entered the forest country: the Linear Pottery culture, clearing the arable land by slash and burn methods. It flourished about 5500–4500 BC, falling entirely within the Atlantic. By the end of the Atlantic, agricultural and pasture lands extended over much of Europe and the once virgin forests were contained within refugia. The end of the Atlantic is signaled by the "Elm decline", a sharp drop in Elm pollen, thought to be the result of human food-producing activities. In the subsequent cooler Sub-Boreal, forested country gave way to open range once more.

Blytt-Sernander

The Blytt-Sernander classification, or sequence, is a series of north European climatic periods or phases based on the study of Danish peat bogs by Axel Blytt (1876) and Rutger Sernander (1908).

The classification was incorporated into a sequence of pollen zones later defined by Lennart von Post, one of the founders of palynology.

Description

Layers in peat were first noticed by Heinrich Dau in 1829. A prize was offered by the Royal Danish Academy of Sciences and Letters to anyone who could explain them. Blytt hypothesized that the darker layers were deposited in drier times; the lighter, in moister times, applying his terms *Atlantic* (warm, moist) and *Boreal* (cool, dry). In 1926 C.A. Webernoticed the sharp boundary horizons, or *grenzhorizont*, in German peat, which matched Blytt's classification. Sernander defined subboreal and subatlantic periods, as well as the late glacial periods. Other scientists have since added other information.

The classification was devised before the development of more accurate dating methods, such as C-14 dating and oxygen isotope ratio cycles. Currently geologists working in different regions are studying sea levels, peat bogs and ice core samples by a variety of methods, with a view towards further verifying and refining the Blytt-Sernander sequence. They find a general correspondence across Eurasia and North America. The fluctuations of climatic change are more complex than Blytt-Sernander period can identify. For example, recent peat core samples at Roskilde Fjord and also Lake Kornerup inDenmark identified 40 to 62 distinguishable layers of pollen, respectively. However, no universally accepted replacement model has been proposed.

Problems

Dating and Calibration

Today the Blytt-Sernander sequence has been substantiated by a wide variety of scientific dating methods, mainly radiocarbon dates obtained from peat. Earlier radiocarbon dates were often left uncalibrated; that is, they were derived by assuming a constant concentration of atmospheric radiocarbon. In fact the atmospheric radiocarbon concentration has varied over time and thus radiocarbon dates need to be calibrated.

Cross-discipline Correlation

The Blytt-Sernander classification has been used as a temporal framework for the archaeological cultures of Europe and America. Some have gone so far as to identify stages of technology in north Europe with specific periods; however, this approach is an oversimplification not generally accepted. There is no reason, for example, why the north Europeans should stop using bronze and start using iron abruptly at the lower boundary of the Subatlantic at 600 BC. In the warm Atlantic period, Denmark was occupied by Mesolithic cultures, rather than Neolithic, notwithstanding the climatic evidence. Moreover, the technology stages vary widely globally.

The Sequence

The Pleistocene phases and approximate calibrated dates are

- Older Dryas stadial, 14,000–13,600 BP (Before Present)
- Allerød interstadial, 13,600–12,900 BP
- Younger Dryas stadial, 12,900–11,640 BP

The Holocene phases are

- Preboreal
- Boreal, cool, dry, rising temperature, 11,500–8,900 BP
- Atlantic, warm, moist, maximum temperature, 8900–5700 BP
- Subboreal, 5700–2600 BP
- Subatlantic, 2600–0 BP

Chemostratigraphy

Chemostratigraphy, or Chemical Stratigraphy, is the study of the variation of chemistry within sedimentary sequences. The name of the field is relatively young, having only come into common usage in the early 1980s, but the basic idea of chemostratigraphy is nearly as old as stratigraphy itself. In some stratigraphic sequences, there is clearly a variation in colour between different strata. Such colour differences often originate from variations in the incorporation of transition metal-containing materials during deposition and lithifaction. Other differences in colour can originate from variations in the organic carbon content of the rock. However, until relatively recently, these variations were not commonly investigated because of the great effort and expense involved in chemical analysis.

However, the development of new analytical techniques for chemical analysis for igneous petrological applications during the latter half of the 20th century, e.g., the electron microprobe, and the development of normal focus X-ray fluorescence for wellsite oil exploration has improved the availability of bulk chemical analysis techniques to the sedimentary geologist, making analysis of the chemical composition of strata increasingly possible. Concurrently, advances in atomic physics stimulated investigations in stable isotope geochemistry. Most relevant to chemostratigraphy in general was the discovery by Harold Urey and Cesare Emiliani in the early 1950s that the oxygen isotope variability in the calcite shells of foraminifera could be used as a proxy for past ocean temperatures.

Thus, chemostratigraphy generally provides two useful types of information to the larger geological community. First, chemostratigraphy can be used to investigate environmental change on the local, regional, and global levels by relating variations in rock chemistry to changes in the environment in which the sediment was deposited. An extreme example of this type of investigation might be the discovery of strata rich in iridium near the boundary between the Cretaceous and Tertiary Systems globally. The high concentration of iridium, which is generally rare in the Earth's crust, is indicative of a large delivery of extraterrestrial material, presumably from a large asteroid impactor during this time. A more prosaic example of chemostratigraphic reconstruction of past conditions might be the use of the Carbon-13/Carbon-12 ratio over geologic time as a proxy for changes in carbon cycle processes at different stages of biological evolution.

Second, regionally or globally correlable chemostratigraphic signals can be found in rocks whose formation time is well-constrained by radionuclide dating of the strata themselves or by strata easily correlated with them, such as a volcanic suite that interrupts nearby strata. However, many sedimentary rocks are much harder to date, because they lack minerals with high concentrations of radionuclides and cannot be correlated with nearly datable sequences. Yet many of these rocks do possess chemostratigraphic signals. Therefore, the correlation between chemostatigraphic signals in conventionally datable and non-datable sequences has extended greatly our understanding of the history of tectonically quiescent regions and of biological organisms that lived in such regions. Chemostratigraphy also has acted as a check on other sub-fields of stratigraphy such as biostratigraphy and magnetostratigraphy.

One interesting aspect of chemostratigraphy from the perspective of stratigraphic formalism is that there is no generally accepted definition or classification scheme for chemostratigraphic units in the North American Stratigraphic Code. Chemostratigraphers generally speak in the terms of radiogenic dates or chronostratigraphic units when referring to particular periods or events of regional or global importance.

Cyclostratigraphy

Cyclostratigraphy is the study of astronomically forced climate cycles within sedimentary successions. Astronomical cycles are variations of the Earth's orbit around the sun due to the gravitational interaction with other masses within the solar system. Due to this cyclicity solar irradiation differs through time on different hemispheres and seasonality is affected. These insolation variations have influence on earth's climate and so on the deposition of sedimentary rocks.

The main orbital cycles are precession with at present main periods of 19- and 23-kyr, obliquity with at present main periods of 41-kyr, and 1,2-Myr, and eccentricity with at present main periods of around 100-kyr, 405-kyr, and 2.4-Myr.

Cyclostratigraphic study of rock records can lead to accurate dating of events in the geological past, to increase understanding of cause and consequences of earth's (climate) history, and to more control on depositional mechanisms of sediments and the acting of sedimentary systems.

$\delta^{13}C$

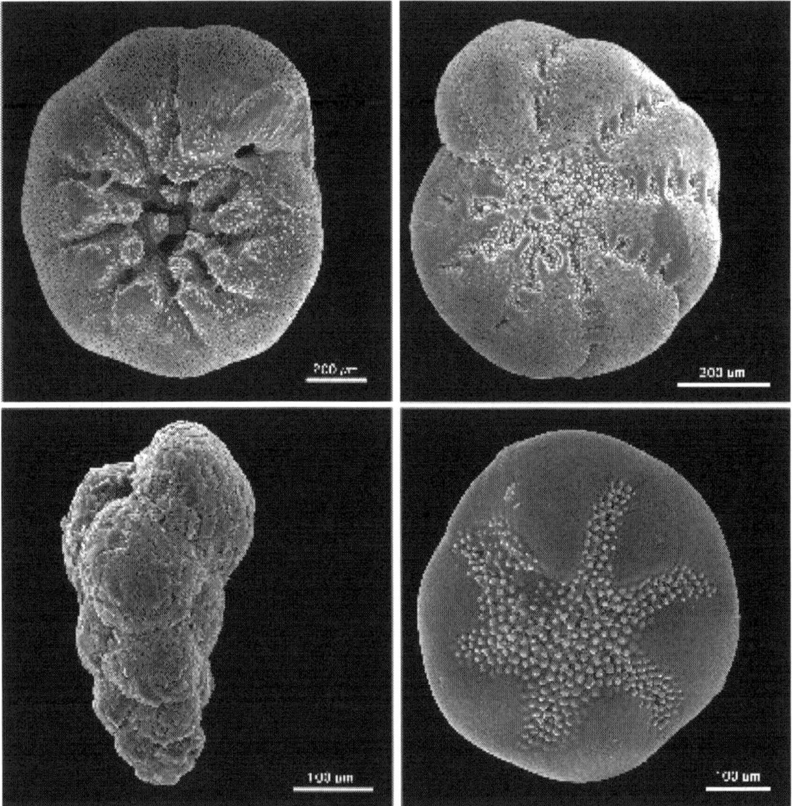

Figure: *Foraminifera samples.*

In geochemistry, paleoclimatology and paleoceanography $\delta^{13}C$ is an isotopic signature, a measure of the ratio of stable isotopes $^{13}C:^{12}C$, reported in parts per thousand (per mil, ‰). $\delta^{13}C$ is pronounced "del C 13".

The definition is, in per mil:

$$\delta^{13}C = \left(\frac{\left(\frac{13_C}{12_C}\right) sample}{\frac{13_C}{12_C} standard} - 1 \right) \times 1000\% \circ$$

where the standard is an established reference material.

$\delta^{13}C$ varies in time as a function of productivity, organic carbon burial and vegetation type.

Reference standard

The standard established for carbon-13 work was the Pee Dee Belemnite or (PDB) and was based on a Cretaceous marine fossil, *Belemnitella americana*, which was from the Pee Dee Formation in South Carolina.

This material had an anomalously high $^{13}C:^{12}C$ ratio (0.0112372), and was established as ^{13}C value of zero. Use of this standard gives most natural material a negative $\delta^{13}C$. The PDB material has been exhausted and the standard replaced by secondary standards.

What Affects $\delta^{13}C$?

Methane has a very light $\delta^{13}C$ signature: biogenic methane of H60‰ thermogenic methane H40‰. The release of large amounts of methane clathrate can impact on global $\delta^{13}C$ values, as at the PETM.

More commonly, the ratio is affected by variations in primary productivity and organic burial. Organisms preferentially take down light 12

C, and have a $\delta^{13}C$ signature of about H25‰, depending on their metabolic pathway. An increase in primary productivity causes a corresponding rise in $\delta^{13}C$ values as more 12

C is locked up in plants. This signal is also a function of the amount of carbon burial; when organic carbon is buried, more 12

C is locked out of the system in sediments than the background ratio (because organic carbon is lighter).

Geologically Significant $\delta^{13}C$ Excursions

C_3 and C_4 plants have different signatures, allowing the importance of C_4 grasses to be detected through time in the $\delta^{13}C$ record. Whereas

C_4 plants have a $\delta^{13}C$ of -16 to -10 ‰, C_3plants have a $\delta^{13}C$ of -33 to -24‰. Mass extinctions are often marked by a negative $\delta^{13}C$ anomaly thought to represent a decrease in primary productivity and release of plant-based carbon. The evolution of large land plants in the late Devonian led to increased organic carbon burial and consequently a rise in $\delta^{13}C$.

Dendroclimatology

Dendroclimatology is the science of determining past climates from trees (primarily properties of the annual tree rings). Tree rings are wider when conditions favour growth, narrower when times are difficult. Other properties of the annual rings, such as maximum latewooddensity (MXD) have been shown to be better proxies than simple ring width. Using tree rings, scientists have estimated many local climates for hundreds to thousands of years previous. By combining multiple tree-ring studies (sometimes with other climate proxy records), scientists have estimated past regional and global climates.

Advantages

Tree rings are especially useful as climate proxies in that they can be well-dated (via matching of the rings from sample to sample, i.e. dendrochronology). This allows extension backwards in time using deceased tree samples, even using samples from buildings or from archeological digs. Another advantage of tree rings is that they are clearly demarked in annual increments, as opposed to other proxy methods such as boreholes.

Furthermore, tree rings respond to multiple climatic effects (temperature, moisture, cloudiness), so that various aspects of climate (not just temperature) can be studied. However, this can be a double-edged sword as discussed in Climate factors.

Limitations

Along with the advantages of dendroclimatology are some limitations: confounding factors, geographic coverage, annular resolution, and collection difficulties. The field has developed various methods to partially adjust for these challenges.

Confounding Factors

There are multiple climate and non-climate factors as well as nonlinear effects that impact tree ring width. Methods to isolate single factors (of interest) include botanical studies to calibrate growth influences and sampling of "limiting stands" (those expected to respond mostly to the variable of interest).

Climate Factors

Climate factors that affect trees include temperature, precipitation, sunlight, and wind. To differentiate among these factors, scientists collect information from "limiting stands". An example of a limiting stand is the upper elevation treeline: here, trees are expected to be more affected by temperature variation (which is "limited") than precipitation variation (which is in excess).

Conversely, lower elevation treelines are expected to be more affected by precipitation changes than temperature variation. This is not a perfect work-around as multiple factors still impact trees even at the "limiting stand", but it helps. In theory, collection of samples from nearby limiting stands of different types (e.g. upper and lower treelines on the same mountain) should allow mathematical solution for multiple climate factors. However, this method is rarely used.

Non-climate Factors

Non-climate factors include soil, tree age, fire, tree-to-tree competition, genetic differences, logging or other human disturbance, herbivore impact (particularly sheep grazing), pest outbreaks, disease, and CO_2 concentration. For factors which vary randomly over space (tree to tree or stand to stand), the best solution is to collect sufficient data (more samples) to compensate for confounding noise. Tree age is corrected for with various statistical methods: either fitting spline curves to the overall tree record or using similar aged trees for comparison over different periods. Careful examination and site selection helps to limit some confounding effects, for example picking sites undisturbed by modern man.

Non-linear Effects

In general, climatologists assume a linear dependence of ring width on the variable of interest (e.g. moisture). However, if the variable changes enough, response may level off or even turn opposite. The home gardner knows that one can underwater or overwater a house plant. In addition, it is possible that interaction effects may occur (for example "temperature times precipitation" may affect growth as well as temperature and precipitation on their own. Here, also, the "limiting stand" helps somewhat to isolate the variable of interest. For instance, at the upper treeline, where the tree is "cold limited", it's unlikely that nonlinear effects of high temperature ("inverted quadratic") will have numerically significant impact on ring width over the course of a growing season.

Botanical Inferences to Correct for Confounding Factors

Botanical studies can help to estimate the impact of confounding variables and in some cases guide corrections for them. These experiments may be either ones where growth variables are all controlled (e.g. in a greenhouse—add ref), partially controlled (e.g. FACE Free Airborne Concentration Enhancement experiments—add ref), or where conditions in nature are monitored. In any case, the important thing is that multiple growth factors are carefully recorded to determine what impacts growth. (Insert Fennoscandanavia paper reference). With this information, ring width response can be more accurately understood and inferences from historic (unmonitored) tree rings become more certain. In concept, this is like the limiting stand principle, but it is more quantitative—like a calibration.

Divergence Problem

The divergence problem is the disagreement between the temperatures measured by the thermometers (instrumental temperatures) on one side and the temperatures reconstructed from the widths of tree rings on the other side, in the northern forests. While the thermometer records indicate a substantial warming trend, many tree rings do not display a corresponding change in their width. A temperature trend extracted from tree rings alone would not show any substantial warming. The temperature graphs calculated in these two ways thus "diverge" from one another since the 1950s, which is the origin of the term. This divergence raises obvious questions of whether other, unrecognized divergences have occurred in the past, prior to the era of thermometers.

Geographic Coverage

Trees do not cover the Earth. Polar and marine climates cannot be estimated from tree rings. In perhumid tropical regions, Australia and southern Africa, trees generally grow all year round and don't show clear annual rings. In some forest areas, the tree growth is too much influenced by multiple factors (no "limiting stand") to allow clear climate reconstruction.

The coverage difficulty is dealt with by acknowledging it and by using other proxies (e.g. ice cores, corals) in difficult areas. In some cases it can be shown that the parameter of interest (temperature, precipitation, etc.) varies similarly from area to area, for example by looking at patterns in the instrumental record. Then one is justified in extending the dendroclimatology inferences to areas where no suitable tree ring samples are obtainable.

Annular Resolution

Tree rings show the impact on growth over an entire growing season. Climate changes deep in the dormant season (winter) will not be recorded. In addition, different times of the growing season may be more important than others (i.e. May versus September) for ring width. However, in general the ring width is used to infer the overall climate change during the corresponding year (an approximation). Another problem is "memory" or autocorrelation. A stressed tree may take a year or two to recover from a hard season. This problem can be dealt with by more complex modelling (a "lag" term in the regression) or by reducing the skill estimates of chronologies.

Collection Difficulties

Tree rings must be obtained from nature, frequently from remote regions. This means that special efforts are needed to map sites properly. In addition, samples must be collected in difficult (often sloping terrain) conditions. Generally, tree rings are collected using a hand-held borer device, that requires skill to get a good sample. The best samples come from felling a tree and sectioning it. However, this requires more danger and does damage to the forest. It may not be allowed in certain areas, particularly with the oldest trees in undisturbed sites (which are the most interesting scientifically). As with all experimentalists, dendroclimatologists must, at times, decide to make the best of imperfect data, rather than resample. This tradeoff is made more difficult, because sample collection (in the field) and analysis (in the lab) may be separated significantly in time and space. These collection challenges mean that data gathering is not as simple or cheap as conventional laboratory science. However, they also give the field's practitioners much enjoyment, working out of doors, with hands on trees and tools.

Other Measurements

Initial work focused on measuring the tree ring width—this is simple to measure and can be related to climate parameters. But the annual growth of the tree leaves other traces. In particular *maximum latewood density* (MXD) is another metric used for estimating environmental variables. It is, however, harder to measure. Other properties (e.g. isotope or chemical trace analysis) have also been tried most notably by L. M. Libby in her 1974 paper "Temperature Dependence of Isotope Ratios in Tree Rings". In theory, multiple measurements on the same ring will allow differentiation of confounding factors (e.g. precipitation and temperature).

However, most studies are still based on ring widths at limiting stands. Measuring radiocarbon concentrations in tree rings has proven to be useful in recreating past sunspot activity, with data now extending back over 11,000 years.

Relationship to Global Warming Study

Tree rings hold the promise of telling us whether 20th century warming is precedented or unprecedented (in last 1000 or so years). The importance of understanding posited global warming from man-made CO_2, has moved dendroclimatology from a sleepy science to a high-profile field. The field has benefitted from increased funding and participation by more researchers. However, the field has also been impacted by the acrimony of the popular debates around global warming. Dendroclimatology is a young science—improvements in methods are being made to squeeze the most insight from tree-ring evidence. While the number of factors affecting growth of a tree ring may seem daunting, there is much good information tied up in tree ring records. Current inferences from tree rings (even if imperfect) are better than knowing nothing about previous climate.

Divergence Problem

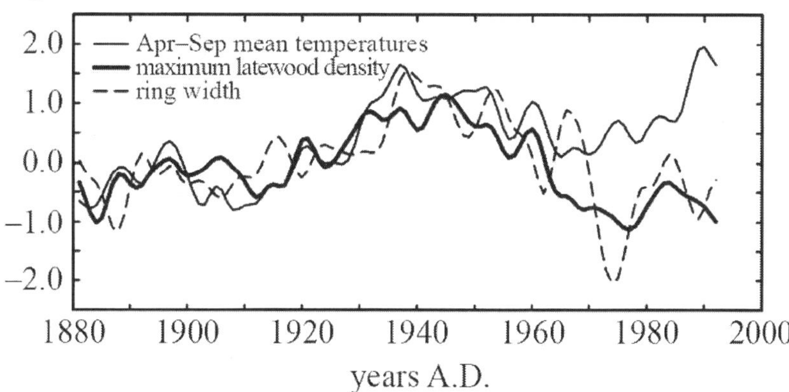

Twenty-year smoothed plots of averaged ring-width (dashed) and tree-ring density (thick line), averaged across all sites, and shown as standardized anomalies from a common base (1881-1940), and compared with equivalent-area averages of mean April–September temperature anomalies (thin solid line). From Briffa 1998 .

The divergence problem is an anomaly from the field of dendroclimatology, the study of past climate through observations of old trees, primarily the properties of their annual growth rings. It is the disagreement between the temperatures measured by thethermometers (instrumental temperatures) and the temperatures reconstructed from the widths of tree rings in the far northern forests.

While the thermometer records indicate a substantial warming trend, many tree rings do not display a corresponding change in their width. A temperature trend extracted from tree rings alone would not show any substantial warming. The temperature graphs calculated in these two ways thus "diverge" from one another since the 1950s, which is the origin of the term.

Samples from southern forests do not exhibit this divergence, though this could be due to paucity of samples, and not all trees in the northern hemisphere do. Divergence is most common in the far northern hemisphere.

Importance

The deviation of some tree ring proxy measurements from the instrumental record since the 1950s raises the question of the reliability of tree ring proxies in the period before the instrumental temperature record. The wide geographic and temporal distribution of well-preserved trees, the solid physical, chemical, and biological basis for their use, and their annual discrimination make dendrochronology particularly important in pre-instrumental climate reconstructions. Tree ring proxies are essentially consistent with other proxy measurements for the period 1600–1950. Before around AD 1600, the uncertainty of temperature reconstructions rises due to the relative paucity of data sets and their limited geographic distribution. At present, these uncertainties are too great to allow conclusion on whether the tree ring record diverges from other proxies during this period.

Possible Explanations

The explanation for the divergence problem is still unclear, but is likely to represent the impact of some other climatic variable that is important to modern northern hemisphere forests but not significant before the 1950s. Rosanne D'Arrigo, senior research scientist at the Tree Ring Lab at Columbia University's Lamont-Doherty Earth Observatory, hypothesises that "beyond a certain threshold level of temperature the trees may become more stressed physiologically, especially if moisture availability does not increase at the same time." Signs suggestive of such stress are visible from space, where satellite pictures show "evidence of browning in some northern vegetation despite recent warming."

Other possible explanations include that the response to recent rapid global warming might be delayed or nonlinear in some fashion.

The divergence might represent changes to other climatic variables to which tree rings are sensitive, such as delayed snowmelt and changes in seasonality. Growth rates could depend more on annual maximum or minimum temperatures, especially in temperature limited growth regions (ie high latitudes and altitudes). Another possible explanation is global dimming due to atmospheric aerosols.

In 2012, Brienen et al. proposed that the divergence problem was largely an artefact of sampling large living trees.

Dole Effect

The Dole effect describes an inequality in the ratio of the heavy isotope O (a 'standard' oxygen atom with two additional neutrons) to the lighter ^{16}O, measured in the atmosphere and seawater. This ratio is usually denoted $\delta^{18}O$. It was noticed in 1935 that air contained more ^{18}O than seawater; this was quantified in 1975 to 23.5‰. The imbalance arises mainly as a result of respiration in plants and in animals. Due to thermodynamics, respiration removes the lighter — hence more reactive — ^{16}O in preference to ^{18}O, increasing the relative amount of ^{18}O in the atmosphere.

The inequality is balanced by photosynthesis. Photosynthesis emits oxygen with the same isotopic composition (i.e. the ratio between ^{18}O and ^{16}O) as the water (H_2O) used in the reaction, which is independent of the atmospheric ratio. Thus when atmospheric ^{18}O levels are high enough, photosynthesis will act as a reducing factor. However, as a complicating factor, the degree of fractionation (i.e. change in isotope ratio) occurring due to photosynthesis is not *entirely* dependent on the water drawn up by the plant, as fractionation can occur as a result of preferential evaporation of $H_2{}^{16}O$ - water bearing lighter oxygen isotopes, and other small but significant processes.

Use of the Dole Effect

Since evaporation causes oceanic and terrestrial waters to have a different ratio of ^{18}O to ^{16}O, the Dole effect will reflect the relevant importances of land-based and marine photosynthesis. The complete removal of land-based productivity would result in a Dole effect shift of -2-3‰ from the current value of 23.5‰. The stability (to within 0.5‰) of the atmospheric ^{18}O to ^{16}O ratio with respect to sea surface waters since the last interglacial (the last 130 000 years), as derived from ice cores, suggests that terrestrial and marine productivity have varied together during this time period.

The Dole effect can also be applied as a tracer in sea water, with slight variations in chemistry being used to track a discrete 'parcel' of water and determine its age.

Dye 3

Dye 3 is an ice core site and previously part of the Distant Early Warning (DEW) line, located at (65°112 N 43°492 W, 2480 masl) in Greenland. As a DEW line base, it was disbanded in years 1990/1991. An ice core is a core sample from the accumulation of snow and ice that has re-crystallized and trapped air bubbles over many years. The composition of these ice cores, especially the presence of hydrogen and oxygen isotopes, provides a picture of the climate at the time. Ice cores contain an abundance of climate information.

Inclusions in the snow, such as wind-blown dust, ash, bubbles of atmospheric gas and radioactive substances, remain in the ice. The variety of climatic proxies is greater than in any other natural recorder of climate, such as tree rings or sediment layers. These include (proxies for) temperature, ocean volume, precipitation, chemistry and gas composition of the loweratmosphere, volcanic eruptions, solar variability, sea-surface productivity, desert extent and forest fires.

Typical ice cores are removed from an ice sheet such as the ice cap internal to Greenland. Greenland is, by area, the world's largest island. The Greenland ice sheet covers about 1.71 million km^2 and contains about 2.6 million km^3 of ice.

Greenland ice sheet

The 'Greenland ice sheet' (Kalaallisut: *Sermersuaq*) is a vast body of ice covering 1.71 million km^2, roughly 80% of the surface of Greenland. It is the second largest ice body in the World, after the Antarctic Ice Sheet. The ice sheet is almost 2,400 kilometers long in a north-south direction, and its greatest width is 1,100 kilometres at a latitude of 77°N, near its northern margin. The mean altitude of the ice is 2,135 metres.

The ice in the current ice sheet is as old as 110,000 years. However, it is generally thought that the Greenland Ice Sheet formed in the late Pliocene or early Pleistocene by coalescence of ice caps and glaciers. It did not develop at all until the late Pliocene, but apparently developed very rapidly with the first continental glaciation.

The ice surface reaches its greatest altitude on two north-south elongated domes, or ridges. The southern dome reaches almost 3,000 metres at latitudes 63°–65°N; the northern dome reaches about 3,290 metres at about latitude 72°N. The crests of both domes are displaced east of the centre line of Greenland. The unconfined ice sheet does not reach the sea along a broad front anywhere in Greenland, so that no large ice shelves occur. On the ice sheet, temperatures are generally substantially lower than elsewhere in Greenland. The lowest mean annual temperatures, about H31 °C (H24 °F), occur on the north-central part of the north dome, and temperatures at the crest of the south dome are about H20 °C (H4 °F). During winter, the ice sheet takes on a strikingly clear blue/green colour. During summer, the top layer of ice melts leaving pockets of air in the ice that makes it look white. Positioned in the Arctic, the Greenland ice sheet is especially vulnerable to global warming. Arctic climate is now rapidly warming.

Distance Early Warning Line

Dye-2 and 3 were among 58 Distance Early Warning (DEW) Line radar stations built by the United States of America (US) between 1955 and 1960 across Alaska, Canada, Greenland and Iceland at a cost of billions of dollars. After extensive studies in late 1957, the US Air Force (USAF) selected sites for two radar stations on the ice cap in southern Greenland. Dye 2 (66°29'30–N 46°18'19–W, 2338 masl) was built approximately 100 miles east of Sondrestrom AB and 90 miles south of the Arctic Circle at an altitude of 7,600 feet.

Dye 3 was located approximately 100 miles east of Dye 2 and slightly south at an elevation of 8,600 feet, contrary to USAF map above. The new radar sites were found to receive from three to four feet of snow each year. The snow was formed into large drifts by winds constantly blowing as much as 100 mph. To overcome this, the Dye sites were elevated approximately 20 feet above the ice cap surface. Dye 3 was completed in 1960.

Dye 3 was disbanded in the years 1990/1991 as a DEW line base.

Greenland Ice Sheet Project (GISP)

The Greenland Ice Sheet Project (GISP) was a decade-long project to drill 20 ice cores in Greenland. GISP involved scientists and funding agencies from Denmark, Switzerland and the United States. Besides the U.S. National Science Foundation, funding was provided by the Swiss National Science Foundation and the Danish Commission for Scientific Research in Greenland. The ice cores provide a proxy archive of temperature and atmospheric constituents that help to understand past climate variations.

Annual field expeditions were carried out to drill intermediate depth cores at various locations on the ice sheet:

- Dye 3 in 1971 to 372 m
- North Site (75°46'N 42°27'W, 2870 masl) in 1972 to 15 m
- North Central (74°37'N 39°36'W) in 1972 to 100 m
- Crête (71°7'N 37°19'W) in 1972 to 15 m
- Milcent (70°18'N 45°35'W, 2410 masl) in 1973 to 398 m
- Dye 2 (66°23'N 46°11'W) in 1973 to 50 m
- Dye 3 in 1973, an intermediate drilling to c. 390 m
- Crête in 1974 to 404.64 m
- Dye 2 in 1974 to 101 m
- Summit (71°17'N 37°56'W, 3212 masl) in 1974 to 31 m
- Dye 3 in 1975 to 95 m
- South Dome (63°33'N 44°36'W, 2850 masl) in 1975 to 80 m
- Hans Tausen (82°30'N 38°20'W, 1270 masl) in 1975 to 60 m
- Dye 3 in 1976 to 93 m
- Hans Tausen in 1976 to 50 m
- Hans Tausen in 1977 to 325 m
- Camp Century (77°10'N 61°8'W, 1885 masl) in 1977 to 49 m
- Dye 2 in 1977 to 84 m
- Camp III (69°43'N 50°8'W) in 1977 to 84 m
- Dye 3 1978 to 90 m
- Camp III in 1978 to 80 m.

"On most of the Greenland ice sheet, however, the annual accumulation rate is considerably higher than 0.2 m ice a^{-1}, and the delta method therefore works thousands of years backwards in time, the only limitation being obliteration of the annual delta cycles by diffusion of the water molecule in the solid ice...."

Delta refers to the changing proportion of oxygen-18 in the different seasonal layers. "The main reason for the seasonal delta variations is that, on its travel to the polar regions, a precipitating air mass is generally cooled more in winter than in summer." "... the annual layer thickness...decreases from 19 cm in 2,000-yr-old ice to 2 cm in 10,000-yr-old ice due to plastic thinning of the annual layers as they sink towards greater depths[10]. volcanic acids in snow layers deposited shortly after a large volcanic eruption can be detected - as elevated specific conductivities measured on melted ice samples8, or as elevated acidities revealed by an electric current through the solid ice..."

Dye 3 Cores

Although available GISP data gathered over the earlier seven years, pointed to north-central Greenland as the optimum site location for the first deep drilling, financial restrictions forced the selection of the logistically convenient Dye-3 location.

Dye 3 1971

Preliminary GISP field work started in 1971 at Dye 3 (65°112 N 43°492 W), where a 372 metre deep, 10.2 cm diameter core was recovered using a Thermal (US) drill type. Three more cores to depths of 90, 93, and 95 m were drilled with different drill types.

Dye 3 1973

For an intermediate drilling c. 390 m, the drill was installed 25 m below the surface at the bottom of the Dye 3 radar station. Some 740 seasonal δ^{18} cycles were counted, indicating that the core reached back to 1231 AD. Evident in this coring was that as melt water seeps through the porous snow, it refreezes somewhere in the cold firn and disturbs the layer sequence.

Dye 3 1975

A second core at Dye 3 was drilled in 1975 with a Shallow (Swiss) drill type to 95 m at 7.6 cm diameter.

Dye 3 1976

A third core at Dye 3 was drilled in 1976 with a Wireline (US) drill type, 10.2 cm diameter, to 93 m.

Dye 3 1978

Another core at Dye 3 was drilled in 1978 using a Shallow (US) drill type, 10.2 cm diameter, to 90 m.

Measurements of [SO_4^{2-}] and [NO_3^-] in firn samples spanning the period 1895-1978 were taken from the Dye 3 1978 core down to 70 m.

Dye 3 1979

In 1979, the initial Dye-3 deep bedrock drilling was started using a 22.2 cm diameter CRREL thermal (US) coring drill to produce an 18 cm diameter access hole, which was cased, to a depth of 77 m. The large diameter casing was inserted over the porous firn zone to contain the drilling fluid.

After working out various logistical and engineering problems related to the development of a more sophisticated drilling rig, drilling to bedrock at Dye 3 began in the summer of 1979 using a new Danish electro-mechanical ice drill yielding a 10.2 cm diameter core. From July to August 1979 using ISTUK, 273 m of core was removed. At the end of the 1980 field season ISTUK had gnawed down to 901 m. In 1981 at a depth of 1785 m dust and conductivity measurements indicated the beginning of ice from the last glaciation. Coring continued and on August 10, 1981, bedrock was reached at a depth of 2038 m. The depth range for the Danish drill was 80–2038 m. The Dye 3 site was a compromise: glaciologically, a higher site on the ice divide with smooth bedrock would have been better; logistically, such a site would have been too remote.

The borehole is 41.5 km east of the local ice divide of the south Greenland ice sheet.

Shear Conditions

The Dye 3 cores were part of the GISP and, at 2037 meters, the final Dye 3 1979 core was the deepest of the 20 ice cores recovered from the Greenland ice sheet. The surface ice velocity is 12.5 ma, 61.2° true. At 500 m above bedrock, the ice velocity is ~10 ma^{H1}, 61.2° true. The ice upstream and downstream from Dye 3 is flowing downhill (-) on ~0.48 % mean slope. The bedrock temperature is H13.22 °C (as of 1984).

Core Continuity

The Dye 3 1979 core is not completely intact and is not undamaged. "Below 600 m, the ice became brittle with increasing depth and badly fractured between 800 and 1,200 m. The physical property of the core progressively improved and below ~1,400 m was of excellent quality." "The deep ice core drilling terminated in August 1981. The ice core is 2035 m long and has a diameter of 10 cm. It was drilled with less than 6° deviation from vertical, and less than 2 m is missing. The deepest 22 m consists of silty ice with an increasing concentration of pebbles downward. In the depth interval 800 to 1400 m the ice was extremely brittle, and even careful handling unavoidably damaged this part of the core, but the rest of the core is in good to excellent condition."

The depth interval 800 to 1400 m would be a period approximately from about two thousand years ago to about five or six thousand years ago.

Melting has been commonplace throughout the Holocene. Summer melting is usually the rule at Dye 3, and there is occasional melting even in north Greenland. All of these meltings disturb the clarity of the annual record to some degree. "An exceptionally warm spell can produce features which extend downwards by percolation, along isolated channels, into the snow of several previous years.

This can happen in regions which generally have little or no melting at the snow surface as exemplified during mid July 1954 in north-west Greenland[4]. Such an event could lead to the conclusion that two or three successive years had abnormally warm summers, whereas all the icing formed during a single period which lasted for several days. The location where melt features will have the greatest climactic significance is high in the percolation facies where summer melting is common but deep percolation is minimal[4]. Dye 3 in southern Greenland (65°11'N; 43°50'W) is such a location."

Counting Annual Layers

As the drill site of Dye 3 receives more than twice as much accumulation as central Greenland, the annual layers are well resolved and relatively thick in the upper parts, making the core ideal for dating the most recent millennia. But, the high accumulation rate has resulted in relatively rapid ice flow (flow-induced layer thinning and diffusion of isotopes), Dye 3 1979 cannot be used for annual layer counting much more than 8 kyr back in time.

Ice Crystal Diameter Distribution

Crystal diameters range from ~0.2 cm at 1900 m from bedrock (depth 137 m) to ~0.42 cm vertical diameter (v) and ~0.55 cm horizontal diameter (h) at 300 m above bedrock (depth 1737 m). However, below 300 m crystal diameter decreases rapidly with increasing dust concentration to a minimum of ~0.05 cm at 200 m above bedrock (depth 1837 m), increasing again linearly to ~0.25 cm v and ~0.3 cm h just above bedrock. Crystal diameters remain approximately constant between 1400 to 300 m above bedrock (depths 637–1737 m), with the largest crystals and the largest distortion (~0.55 cm v and ~0.7 cm h) occurring at 1100 m above bedrock (depth 937 m).

The brittle zone mentioned above under "Core continuity" corresponds in Dye 3 1979 with the steady state grain size (crystal size) from ~637 - ~1737 m depth range. This is also the Holocene climatic optimum period.

Beryllium 10 Variations

As of 1998 the only long record available for ^{10}Be is from Dye 3 1979. Questions were raised whether all parts of the Dye 3 1979 record reflect the sun activity or are affected by climatic and/or ice dynamics.

Dust Concentration

The dust concentration has a peak of ~3 mg/kg at 200 m above bedrock (depth 1837 m), second only to the silty ice (>20 mg/kg) of the bottom 25 m, which has a very high deformation rate.

Ice Ages

The uppermost 1780 m is considered Holocene ice, and the lower portion is considered as deposited during the Wisconsin period. From the δ^{18} O profile of the Dye 3 core it is relatively easy to differentiate the post-glacial climatic optimum, portions thereof and earlier: the Pre-Boreal transition, the Allerød, Bølling, Younger Dryas, and Oldest Dryas.

In the Dye 3 1979 oxygen isotope record, the Older Dryas appears as a downward peak establishing a small, low-intensity gap between the Bølling and the Allerød. During the transition from the Younger Dryas to the Pre-Boreal, the South Greenland temperature increased by 15 °C in 50 years. At the beginning of this same transition the deuterium excess and dust concentration shifted to lower levels in less than 20 years. The post-glacial climatic optimum lasted from ~9000-4000 yrs B.P. as determined from Dye 3 1979 and Camp Century 1963 δ^{18} O profiles. Both Dye 3 1979 and Camp Century 1963 cores exhibit the 8.2 ka event and the boundary event separating Holocene I from Holocene II.

Fossils

Samples from the base of the 2 km deep Dye 3 1979 and the 3 km deep GRIP cores revealed that high-altitude southern Greenland has been inhabited by a diverse array of conifer trees and insects within the past million years.

Dye 3 1988

Ellen Mosley-Thompson led a 3-man glaciological team to drill an intermediate depth core at Dye 3, Greenland.

Comparison with other Greenland ice cores

A graphical description of changes in temperature in Greenland from AD 500 – 1990 based on analysis of the deep ice core from Greenland and some historical events. The annual temperature changes are shown vertical in ÚC. The numbers are to be read horizontal:

1. From AD 700 to 750 people belonging to the Late Dorset Culture move into the area around Smith Sound, Ellesmere Island and Greenland north of Thule.

2. Norse settlement of Iceland starts in the second half of the 9th century.
3. Norse settlement of Greenland starts just before the year 1000.
4. Thule Inuit move into northern Greenland in the 12th century.
5. Late Dorset culture disappears from Greenland in the second half of the 13th century.
6. The Western Settlement disappears in mid 14th century.
7. In 1408 is the Marriage in Hvalsey, the last known written document on the Norse in Greenland.
8. The Eastern Settlement disappears in mid 15th century.
9. John Cabot is the first European in the post-Iceland era to visit Labrador - Newfoundland in 1497.
10. "Little Ice Age" from ca 1600 to mid 18th century.
11. The Norwegian priest, Hans Egede, arrives in Greenland in 1721.

To investigate the possibility of climatic cooling, scientists drilled into the Greenland ice caps to obtain core samples. The oxygen isotopes from the ice caps suggested that the Medieval Warm Period had caused a relatively milder climate in Greenland, lasting from roughly 800 to 1200. However from 1300 or so the climate began to cool. By 1420, we know that the "Little Ice Age" had reached intense levels in Greenland.

For most of the arctic ice cores up to 1987, regions of the core with high dust concentrations correlate well with the ice having high deformation rates and small crystal diameters, in both Holocene and Wisconsin ice.

Camp Century 1963

The Camp Century, Greenland, ice core (cored from 1963–1966) is 1390 m deep and contains climatic oscillations with periods of 120, 940, and 13,000 years.

Counting Annual Layers

> *"Thus in principle dating of the Camp Century ice core by counting annual layers is possible to about the 1,060 m depth, corresponding to 8,300 yr BP according to the time scale which we shall adopt." "It may be necessary, however, to apply a depth dependent correction to account for 'lost' annual oscillations. Even during firnification seasonal d-oscillations in years with unusually low*

accumulation may disappear due to mass exchange. Unfortunately, the physical condition (broken or missing pieces) of the Camp Century ice core precludes continuous measurement of seasonal isotope variations for the purpose of dating from the surface downward."

Crête 1972

The Crête core was drilled in central Greenland (1974) and reached a depth of 404.64 meters, extending back only about fifteen centuries.

Milcent 1973

"The first core drilled at Station Milcent in central Greenland covers the past 780 years." Milcent core was drilled at 70.3°N, 44.6°W, 2410 masl. The Milcent core (398 m) was 12.4 cm in diameter, using a Thermal (US) drill type, in 1973.

Ice Ages

The Milcent core record only goes back to AD 1174 (Holocene) due to the high accumulation rates.

Renland 1985

The Renland ice core was drilled in 1985. The Renland ice core from East Greenland apparently covers a full glacial cycle from the Holocene into the previous Eemian interglacial. The Renland ice core is 325 m long.

From the delta-profile, the Renland ice cap in the Scoresbysund Fiord has always been separated from the inland ice, yet all the delta-leaps revealed in the Camp Century 1963 core recurred in the Renland ice core.

Inclusions in the Ice

The Renland core is noted for apparently containing the first Northern Hemisphere record of methanesulfonate (MSA), and having the first continuous record of non-seasalt sulfate.

The Renland core is also the first to provide a continuous record of ammonium (NH_4) apparently through the whole glacial period.

The distribution of ^{10}Be in the top 40 m of the Renland ice core has been reported and corroborates the ^{10}Be cyclic fluctuation pattern from Dye 3.

Ice Ages

The Renland core apparently contains ice from the Eemian onward.

Grip 1989

GRIP successfully drilled a 3028 metre ice core to the bed of the Greenland Ice Sheet at Summit, Central Greenland from 1989 to 1992 at 72°352 N 37°382 W, 3238 masl.

Inclusions in the Ice

Eight ash layers have been identified in the central Greenland ice core GRIP. Four of the ash layers (Ash Zones I and II, Saksunarvatn and the Settlement layer) originating in Iceland have been identified in GRIP by comparison of chemical composition of glass shards from the ash.

The other four have not been correlated with known ash deposits. The Saksunarvatn tephra via radiocarbon dating is ca 10,200 years BP.

Gisp2 1989

The follow-up U.S. GISP2 project drilled at a glaciologically better location on the summit (72°36'N, 38°30'W, 3200 masl). This hit bedrock (and drilled another 1.55 m into bedrock) on July 1, 1993 after five years of drilling. European scientists produced a parallel core in the GRIP project. GISP2 produced an ice core 3053.44 metres in depth, the deepest ice core recovered in the world at the time. The GRIP site was 30 km to the east of GISP2. "Down to a depth of 2790 m in GISP2 (corresponding to an age of about 110 kyr B.P.), the GISP2 and GRIP records are nearly identical in shape and in many of the details."

Visual Stratigraphy

The GISP2 time scale is based on counting annual layers primarily by visual stratigraphy. The isotopic temperature records show 23 interstadial events correlateable between the GRIP and GISP2 records between 110 and 15 kyr B.P. Ice in both cores below 2790 m depth (records prior to 110 kyr B.P.) shows evidence of folding or tilting in structures too large to be fully observed in a single core.

The bulk of the GISP2 ice core is archived at the National Ice Core Laboratory in Lakewood, Colorado, United States.

North Greenland Ice Core Project 1996

The drilling site of the North Greenland Ice Core Project (NGRIP) is near the centre of Greenland (75.1 N, 42.32 W, 2917 m, ice thickness 3085). Drilling began in 1999 and was completed at bedrock in on July 17, 2003. The NGRIP site was chosen to extract a long and undisturbed record stretching into the last glacial, and it was apparently successful.

Unusually, there is melting at the bottom of the NGRIP core - believed to be due to a high geothermal heat flux locally. This has the advantage that the bottom layers are less compressed by thinning than they would otherwise be: NGRIP annual layers at 105 kyr age are 1.1 cm thick, twice the GRIP thicknesses at equal age.

Shear Conditions

The site was chosen for a flat basal topography to avoid the flow distortions that render the bottom of the GRIP and GISP cores unreliable.

Ice Crystal Diameter Distribution

In the upper 80 m of the ice sheet, the firn or the snow gradually compacts to a close packing of ice crystals of typical sizes 1 to 5 mm. Crystal size distributions were obtained from fifteen vertical thin sections of 20 cm x 10 cm (height x width) and a thickness of 0.4 ±0.1 mm of ice evenly distributed in the depth interval 115 – 880 m. Peak sizes with depth were ~1.9 mm 115 m, ~2.2 mm 165 m, ~2.8 mm 220 m, ~3.0 mm 330 m, ~3.2 mm 440 m, ~3.3 mm 605 m, whereas mean sizes with depth were ~1.8 115 m, ~2.2 mm 165 m, ~2.4 mm 220 m, ~2.8 mm ~270 m, ~2.75 mm 330 m, ~2.6 mm ~370 m, ~2.9 mm 440 m, ~2.8 mm ~490 m, ~2.9 mm ~540 m, ~2.9 mm 605 m, ~3.0 ~660 m, ~3.2 mm ~720 m, ~2.9 mm ~770 m, ~2.7 mm ~820 m, ~2.8 mm 880 m. And, here again as with Dye 3, steady state in grain growth was reached and continued through the post-glacial climatic optimum.

The size distribution of ice crystals changes with depth and approaches the Normal Grain Growth law via competing mechanisms of fragmentation (producing smaller polygonal grains) and grain boundary diffusion (producing larger, vertically compressed, horizontally expanded grains). Although some of the peaks for the deeper distributions appear to be slightly greater, the predicted steady state average grain size is 2.9±0.1 mm.

Ice Ages

The NGRIP record helps to resolve a problem with the GRIP record - the unreliability of the Eemian Stage portion of the record. NGRIP covers 5 kyr of the Eemian, and shows that temperatures then were roughly as stable as the pre-industrial Holocene temperatures were. This is confirmed by sediment cores, in particular MD95-2042.

Fossils

In 2003, NGRIP recovered what seem to be plant remnants nearly two miles below the surface, and they may be several million years old.

"Several of the pieces look very much like blades of grass or pine needles," said University of Colorado at Boulder geological sciences Professor James White, an NGRIP principal investigator. "If confirmed, this will be the first organic material ever recovered from a deep ice-core drilling project," he said.

100,000-year Problem

The 100,000-year problem is a discrepancy between past temperatures and the amount of incoming solar radiation, orinsolation. The former rises and falls according to the strength of radiation from the sun, the distance from the earth to the sun, and the tilt of the Earth's poles. However, the ice-age cycle, which grows and shrinks periodically on a 100,000-year (100 ka) timescale, does not correlate well with any of these factors.

Due to variations in the Earth's orbit, the amount of insolation varies with periods of around 21,000, 40,000, 100,000, and 400,000 years. Variations in the amount of solar heating drive changes in the climate of the Earth, and are recognised as a key factor in the timing of initiation and termination of ice ages. Spectral analysis shows that the most powerful climate response is at 100,000-year period, but the orbital forcing at this period is small.

Reconstructing Past Climate

Past climate data—especially temperature—can be readily inferred from sedimentary evidence, although not with the accuracy that instruments can measure current temperatures. Perhaps the most useful indicator of past climate is the fractionation of oxygen isotopes, denoted $\delta^{18}O$. This fractionation is controlled mainly by the amount of water locked up in ice and the absolute temperature of the planet, and has allowed a timescale of marine isotope stages to be constructed.

Comparing the Records

The $\delta^{18}O$ record of air (in the Vostok ice core) and marine sediments has been compared with estimates of solar insolation, which should affect both temperature and ice volume. Nicholas Shackleton orbitally tuned the Antarctic ice core air $\delta^{18}O$ (i.e. he matched the signal to this assumed forcing), and used spectral analysis to identify and subtract the component of the record that in this interpretation could be attributed to a linear (directly proportional) response to the orbital forcing. The residual signal (the remainder), when compared with the residual from a similarly retuned marine core isotope record, allowed him to estimate the proportion of the signal that was attributable to ice volume, with the rest (having attempted to allow for the Dole effect) being attributed to temperature changes in the deep water. The 100,000-year component of ice volume variation was found to match sea level records based on coral age determinations, and to lag orbital eccentricity by several thousand years, as would be expected if orbital eccentricity were the pacing mechanism. Strong non-linear "jumps" in the record appear at deglaciations, although the 100,000-year periodicity was not the strongest periodicity in this "pure" ice volume record. The separate deep sea temperature record was found to vary directly in phase with orbital eccentricity, as did Antarctic temperature and CO_2; so eccentricity appears to exert a geologically immediate effect on air temperatures, deep sea temperatures, and atmospheric carbon dioxide concentrations. Shackleton concluded: "The effect of orbital eccentricity probably enters the paleoclimatic record through an influence on the concentration of atmospheric CO_2". The mechanism causing these cyclic temperature changes remains at the heart of the 100,000-year problem.

Solutions to the Problem

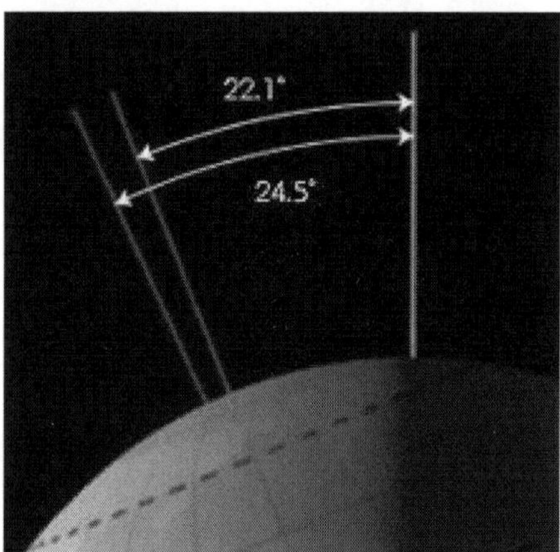

Figure: *The effect of obliquity variations may, in concert with precession, be amplified by orbital inclination.*

As the 100,000-year periodicity only dominates the climate of the past million years, there is insufficient information to separate the component frequencies of eccentricity using spectral analysis, making the reliable detection of significant longer-term trends more difficult, although the spectral analysis of much longer palaeoclimate records, such as the Lisiecki and Raymo stack of marine cores and James Zachos' composite isotopic record, helps to put the last million years in longer term context. Hence there is still no sheer consensus on the mechanism responsible for the 100ka periodicity—but there are several popular contenders. The mechanism may be internal to the Earth system. The Earth's climate system may have a natural resonance frequency of 100ka; that is to say, feedback processes within the climate automatically produce a 100ka effect, much as a bell naturally rings at a certain pitch. Opponents to this claim point out that the resonance would have to have developed 1 million years ago, as a 100ka periodicity was weak to non-existent for the preceding 2 million years. This is not infeasible—continental drift and sea floor spreading rate change have been postulated as possible causes of such a change. Free oscillations of components of the Earth system have been considered as a cause, but too few Earth systems have a thermal inertia on a thousand-year timescale for any long-term changes to accumulate.

Orbital inclination has a 100ka periodicity, while eccentricity's 95 and 125ka periods combine to give a 100ka effect. While it is possible that the less significant, and originally overlooked, inclination variability has a deep effect on climate, the eccentricity only modifies insolation by a small amount: 1–2% of the shift caused by the 21,000-year precession and 41,000-year obliquity cycles. Such a big impact would therefore be disproportionate in comparison to other cycles. One possible mechanism suggested to account for this was the passage of Earth through regions of cosmic dust. Our eccentric orbit would take us through dusty clouds in space, which would act to occlude some of the incoming radiation, shadowing the Earth.

In such a scenario, the abundance of the isotope He, produced y solar rays splitting gases in the upper atmosphere, would be expected to decrease—and initial investigations did indeed find such a drop in He abundance. However, there is still the possibility that the 100ka eccentricity cycle acts as a "pacemaker" to the system, amplifying the effect of precession and obliquity cycles at key moments, pushing a system usually in an accumulatory state "over the brink" into a swift melting phase, by providing the lightest of taps.

A similar suggestion holds the 21,636-year precession cycles solely responsible. Ice ages are characterized by the slow buildup of ice volume, followed by relatively swift melting phases. It is possible that ice built up over several precession cycles, only melting after four or five such cycles.

A mechanism that may account for periodic fluctuations in solar luminosity has also been proposed as an explanation. Diffusion waves occurring within the sun can be modelled in such a way that they explain the observed climatic shifts on earth. However, the He signal again appears to contradict this finding. The Dole effect describes trends in δO arising from trends in the relative importance of land-dwelling and oceanic photosynthesizers. Such a variation is a plausible perpetrator of the phenomenon, but only really passes the buck: what changed the importances of land- and sea-based photosynthesis?

The recovery of higher-resolution ice cores spanning more of the past 1,000,000 years by the ongoing EPICA project may help to shed more light on the matter. A new, high-precision dating method developed by the team allows better correlation of the various factors involved and puts the ice core chronologies on a stronger temporal footing, endorsing the traditional Milankovitch hypothesis, that climate variations are controlled by insolation in the northern hemisphere. The new chronology is inconsistent with the "inclination" theory of the 100,000-year cycle.

The establishment of leads and lags against different orbital forcing components with this method—which uses the direct insolation control over nitrogen-oxygen ratios in ice core bubbles—is in principle a great improvement in the temporal resolution of these records and another significant validation of the Milankovitch hypothesis.

8.2 Kiloyear Event

The 8.2 kiloyear event is the term that climatologists have adopted for a sudden decrease in global temperatures that occurred approximately 8,200 years before the present, or c. 6,200 BCE, and which lasted for the next two to four centuries. Milder than the Younger Dryas cold spell that preceded it, but more severe than the Little Ice Age that would follow, the 8.2 kiloyear cooling was a significant exception to general trends of the Holocene climatic optimum. During the event, atmospheric methane concentration decreased by 80 ppb or 15% emission reduction by cooling and drying at a hemispheric scale.

A rapid cooling around 6200 BCE was first identified by Swiss botanist Heinrich Zoller in 1960, who named the event *Misox oscillation* (for the Val Mesolcina). It is also known as *Finse event* in Norway. Bond *et al.* argued that the origin of the 8.2 kiloyear event is linked to a 1,500-year climate cycle; it correlates with Bond event 5.

The strongest evidence for the event comes from the North Atlantic region; the disruption in climate shows clearly in Greenland ice coresand in sedimentary and other records of the temporal and tropical North Atlantic. It is less evident in ice cores from Antarctica and in South American indices. The effects of the cold snap were global, however, most notably in changes in sea level during the relevant era. The 8.2 Ka cooling event may have been caused by a large meltwater pulse from the final collapse of the Laurentide ice sheet of northeastern North America—most likely when the glacial lakes Ojibway and Agassiz suddenly drained into the North Atlantic Ocean. (The same type of action produced the Missoula floods that created the Channeled scablands of the Columbia River basin.) The meltwater pulse may have affected the North Atlantic thermohaline circulation, reducing northward heat transport in the Atlantic and causing significant circum-North Atlantic cooling.

Estimates of the cooling vary and depend somewhat on the interpretation of the proxy data, but drops of around 1 to 5 °C (1 to 11 °F) have been reported. In Greenland, the event started at 8175 Before Present, and the cooling was 3.3 °C (decadal average) in less than ~20 years, and the coldest period lasted for about 60 years, and the total duration was about 150 years. Further afield, some tropical records report a 3 °C (5 °F) cooling from cores drilled into an ancient coral reef in Indonesia. The event also caused a global CO_2 decline of ~ 25 ppm over ~ 300 years. However, the dating and interpretation of this and other tropical sites are more ambiguous than the North Atlantic sites.

Drier conditions were notable in North Africa, while East Africa suffered five centuries of general drought. In West Asia and especially Mesopotamia, the 8.2ky event was a three-hundred year aridification and cooling episode, which provided the natural force for Mesopotamian irrigation agriculture and surplus production that were essential for the earliest class-formation and urban life. However multi-centennial changes around the same period are difficult to link specifically to the approximately 100-year abrupt event as recorded most clearly in the Greenland ice cores.

The initial meltwater pulse has caused between 0.5 and 4 meters of sea-level rise. Based on estimates of lake volume and decaying ice cap size, values of 0.4–1.2 meters (1–4 ft) circulate. Based on sea-level data from below modern deltas 2–4 meters (6–12 ft) of near-instantaneous rise is estimated, recorded superimposed on background 'normal' post-glacial sea-level rise.

Meltwater pulse sea level rise was experienced fully at great distance from the release area. Gravity and rebound effects associated to the shifting of watermasses mean that the sea-level fingerprint is smaller in areas closer to the Hudson Bay. The Mississippi delta records ~20%, NW Europe records ~70% and Asia records ~105% of the global averaged amount. The cooling of the 8200 event was a temporary feature, the sea-level rise of the meltwater pulse was permanent. In 2003, the Office of Net Assessment at the United States Department of Defence was commissioned to produce a study on the likely and potential effects of a modern climate change. The study, conducted under ONA head Andrew Marshall, modelled its prospective climate change on the 8.2 kiloyear event, precisely because it was the middle alternative between the Younger Dryas and the Little Ice Age.

Airborne Fraction

The airborne fraction is a scaling factor defined as the ratio of the annual increase in atmospheric CO_2 to the CO_2 emissions from anthropogenic sources. It represents the proportion of human emitted CO_2 that remains in the atmosphere. The fraction averages about 45%, meaning that approximately half the human-emitted CO_2 is absorbed by ocean and land surfaces. There is some evidence for a recent increase in airborne fraction, which would imply a faster increase in atmospheric CO_2 for a given rate of human fossil-fuel burning. However, other sources suggest that the "fraction of carbon dioxide has not increased either during the past 150 years or during the most recent five decades".

7

Greenhouse Effect

The greenhouse effect is a process by which thermal radiation from a planetary surface is absorbed by atmospheric greenhouse gases, and is re-radiated in all directions. Since part of this re-radiation is back towards the surface and the lower atmosphere, it results in an elevation of the average surface temperature above what it would be in the absence of the gases.

Solar radiation at the frequencies of visible light largely passes through the atmosphere to warm the planetary surface, which then emits this energy at the lower frequencies of infrared thermal radiation. Infrared radiation is absorbed by greenhouse gases, which in turn re-radiate much of the energy to the surface and lower atmosphere. The mechanism is named after the effect of solar radiation passing through glass and warming a greenhouse, but the way it retains heat is fundamentally different as a greenhouse works by reducing airflow, isolating the warm air inside the structure so that heat is not lost by convection.

The existence of the greenhouse effect was argued for by Joseph Fourier in 1824. The argument and the evidence was further strengthened by Claude Pouillet in 1827 and 1838, and reasoned from experimental observations by John Tyndall in 1859, and more fully quantified by Svante Arrhenius in 1896. If an ideal thermally conductive blackbody was the same distance from the Sun as the Earth is, it would have a temperature of about 5.3 °C. However, since the Earth reflects about 30% of the incoming sunlight, the planet's effective temperature (the temperature of a blackbody that would emit the same amount of radiation) is about e18 °C, about 33°C below the actual surface temperature of about 14 °C. The mechanism that produces this difference between the actual surface temperature and the effective temperature is due to the atmosphere and is known as the greenhouse effect.

Earth's natural greenhouse effect makes life as we know it possible. However, human activities, primarily the burning of fossil fuels and clearing of forests, have intensified the natural greenhouse effect, causing global warming.

Basic Mechanism

The Earth receives energy from the Sun in the form UV, visible, and near IR radiation, most of which passes through the atmosphere without being absorbed. Of the total amount of energy available at the top of the atmosphere (TOA), about 50% is absorbed at the Earth's surface. Because it is warm, the surface radiates far IR thermal radiation that consists of wavelengths that are predominantly much longer than the wavelengths that were absorbed (the overlap between the incident solar spectrum and the terrestrial thermal spectrum is small enough to be neglected for most purposes). Most of this thermal radiation is absorbed by the atmosphere and re-radiated both upwards and downwards; that radiated downwards is absorbed by the Earth's surface. This trapping of long-wavelength thermal radiation leads to a higher equilibrium temperature than if the atmosphere were absent.

This highly simplified picture of the basic mechanism needs to be qualified in a number of ways, none of which affect the fundamental process.

The solar radiation spectrum for direct light at both the top of the Earth's atmosphere and at sea level

- The incoming radiation from the Sun is mostly in the form of visible light and nearby wavelengths, largely in the range 0.2–4 μm, corresponding to the Sun's radiative temperature of 6,000 K. Almost half the radiation is in the form of "visible" light, which our eyes are adapted to use.
- About 50% of the Sun's energy is absorbed at the Earth's surface and the rest is reflected or absorbed by the atmosphere. The reflection of light back into space—largely by clouds—does not much affect the basic mechanism; this light, effectively, is lost to the system.
- The absorbed energy warms the surface. Simple presentations of the greenhouse effect, such as the idealized greenhouse model, show this heat being lost as thermal radiation. The reality is more complex: the atmosphere near the surface is

largely opaque to thermal radiation (with important exceptions for "window" bands), and most heat loss from the surface is by sensible heat and latent heat transport.

Radiative energy losses become increasingly important higher in the atmosphere largely because of the decreasing concentration of water vapor, an important greenhouse gas. It is more realistic to think of the greenhouse effect as applying to a "surface" in the mid-troposphere, which is effectively coupled to the surface by a lapse rate.

- The simple picture assumes a steady state. In the real world there is the diurnal cycle as well as seasonal cycles and weather. Solar heating only applies during daytime. During the night, the atmosphere cools somewhat, but not greatly, because its emissivity is low, and during the day the atmosphere warms. Diurnal temperature changes decrease with height in the atmosphere.
- Within the region where radiative effects are important the description given by the idealized greenhouse model becomes realistic: The surface of the Earth, warmed to a temperature around 255 K, radiates long-wavelength, infrared heat in the range 4–100 μm. At these wavelengths, greenhouse gases that were largely transparent to incoming solar radiation are more absorbent. Each layer of atmosphere with greenhouses gases absorbs some of the heat being radiated upwards from lower layers. It re-radiates in all directions, both upwards and downwards; in equilibrium (by definition) the same amount as it has absorbed. This results in more warmth below. Increasing the concentration of the gases increases the amount of absorption and re-radiation, and thereby further warms the layers and ultimately the surface below.
- Greenhouse gases—including most diatomic gases with two different atoms (such as carbon monoxide, CO) and all gases with three or more atoms—are able to absorb and emit infrared radiation. Though more than 99% of the dry atmosphere is IR transparent (because the main constituents—N_2, O_2, and Ar—are not able to directly absorb or emit infrared radiation), intermolecular collisions cause the energy absorbed and emitted by the greenhouse gases to be shared with the other, non-IR-active, gases.

Greenhouse Gases

By their percentage contribution to the greenhouse effect on Earth the four major gases are:

- water vapor, 36–70%
- carbon dioxide, 9–26%
- methane, 4–9%
- ozone, 3–7%

The major non-gas contributor to the Earth's greenhouse effect, clouds, also absorb and emit infrared radiation and thus have an effect on radiative properties of the atmosphere.

Role in Climate Change

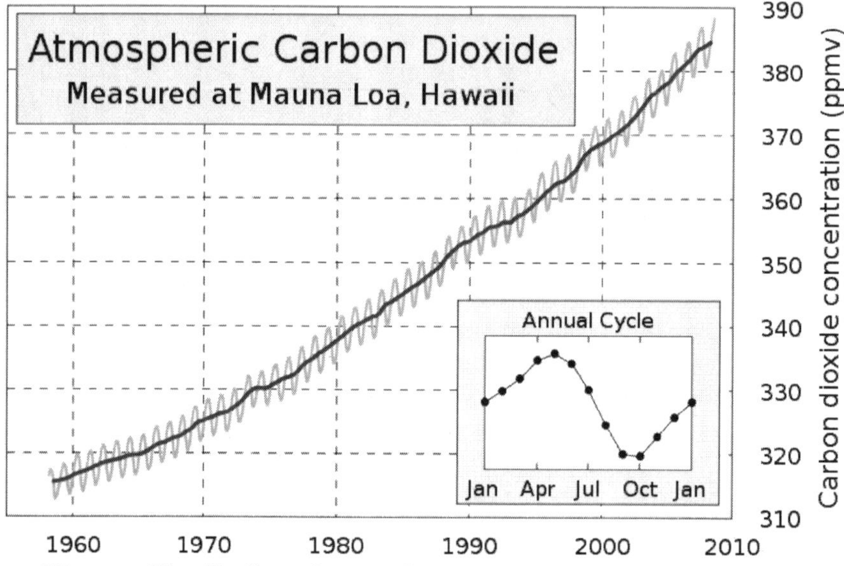

Figure: *The Keeling Curve of atmospheric CO_2 concentrations measured at Mauna Loa Observatory.*

Strengthening of the greenhouse effect through human activities is known as the enhanced (or anthropogenic) greenhouse effect. This increase in radiative forcing from human activity is attributable mainly to increased atmospheric carbon dioxide levels. According to the latest Assessment Report from the Intergovernmental Panel on Climate Change,

"*most of the observed increase in globally averaged temperatures since the mid-20th century is very likely due to the observed increase in anthropogenic greenhouse gas concentrations*".

CO_2 is produced by fossil fuel burning and other activities such as cement production and tropical deforestation. Measurements of CO_2 from the Mauna Loa observatory show that concentrations have increased from about 313 ppm in 1960 to about 389 ppm in 2010. The current observed amount of CO_2 exceeds the geological record maxima (~300 ppm) from ice core data. The effect of combustion-produced carbon dioxide on the global climate, a special case of the greenhouse effect first described in 1896 by Svante Arrhenius, has also been called the Callendar effect.

Over the past 800,000 years, ice core data shows that carbon dioxide has varied from values as low as 180 parts per million (ppm) to the pre-industrial level of 270ppm. Paleoclimatologists consider variations in carbon dioxide concentration to be a fundamental factor influencing climate variations over this time scale.

Real Greenhouses

The "greenhouse effect" of the atmosphere is named by analogy to greenhouses which get warmer in sunlight, but the mechanism by which the atmosphere retains heat is different. A greenhouse works primarily by preventing absorbed heat from leaving the structure through convection, i.e. sensible heat transport.

The greenhouse effect heats the earth because greenhouse gases absorb outgoing radiative energy and re-emit some of it back towards earth. A greenhouse is built of any material that passes sunlight, usually glass, or plastic. It mainly heats up because the Sun warms the ground inside, which then warms the air in the greenhouse.

The air continues to heat because it is confined within the greenhouse, unlike the environment outside the greenhouse where warm air near the surface rises and mixes with cooler air aloft. This can be demonstrated by opening a small window near the roof of a greenhouse: the temperature will drop considerably.

It has also been demonstrated experimentally (R. W. Wood, 1909) that a "greenhouse" with a cover of rock salt (which is transparent to infra red) heats up an enclosure similarly to one with a glass cover. Thus greenhouses work primarily by preventing convective cooling. In the greenhouse effect, rather than retaining (sensible) heat by physically preventing movement of the air, greenhouse gases act to warm the Earth by re-radiating some of the energy back towards the surface. This process may exist in real greenhouses, but is comparatively unimportant there.

Bodies Other Than Earth

In our solar system, Mars, Venus, and the moon Titan also exhibit greenhouse effects. Titan has an anti-greenhouse effect, in that its atmosphere absorbs solar radiation but is relatively transparent to infrared radiation. Pluto also exhibits behaviour superficially similar to the anti-greenhouse effect. A runaway greenhouse effect occurs if positive feedbacks lead to the evaporation of all greenhouse gases into the atmosphere. A runaway greenhouse effect involving carbon dioxide and water vapor is thought to have occurred on Venus. List of periods and events in climate history. Knowledge of precise climatic events decreases as the record goes further back in time. Some notable climate events known to Paleoclimatology are listed here. The Timeline of glaciation covers ice ages specifically, which tend to have their own names for phases, often with different names used for different parts of the world. The names for earlier periods and events come from geology and paleontology. The Marine isotope stages (MIS) are often used to express dating within the Quaternary.

Pleistocene

All dates are approximate. "(B-S)" means this is one of the periods from the Blytt-Sernander sequence, originally based on studies of Danish peat bogs.

- 120,000 - 90,000 BP Abbassia Pluvial wet in North Africa
- 110,000 - 10,000 BP Last glacial period, not to be confused with the Last Glacial Maximum or Late Glacial Maximum below
  - 50,000 - 30,000 BP Mousterian Pluvial wet in North Africa
  - 26,500 - 19,000–20,000 BP Last Glacial Maximum, what is often meant in popular usage by "Last Ice Age"
  - 16,000 - 13,000 BC Oldest Dryas cold, begins slowly and ends sharply (B-S)
  - 12,700 BC Antarctic Cold Reversal warmer Antarctic, sea levels rise
  - 12,400 BC Bølling oscillation warm and wet in the North Atlantic, begins the Bølling-Allerød period (B-S)
  - 12,400 - 11,500 BC (much discussed) Older Dryas cold, interrupts warm period for some centuries (B-S)
  - 12,000 - 11,000 BC Allerød oscillation warm & moist (B-S)
  - 11,400 - 9,500 BC Huelmo/Mascardi Cold Reversal cold in Southern Hemisphere

- o 11,000-8,000 BC Late Glacial Maximum, or Tardiglacial (definitions vary)
- o 10,800 - 9,500 BC Younger Dryas sudden cold and dry period in Northern Hemisphere (B-S)

Holocene

All dates are BC (BCE) and approximate. "(B-S)" means this is one of the periods from the Blytt-Sernander sequence, originally based on studies of Danish peat bogs.

- from 10,000 Holocene glacial retreat, the present Holocene or Postglacial period begins
- 9,400 BC Pre-Boreal sharp rise in temperature over 50 years (B-S), precedes Boreal
- 8,500 - 6,900 BC Boreal (B-S), rising sea levels, forest replaces tundra in northern Europe
- 7,500 - 3,900 Neolithic Subpluvial in North Africa, wet
- 7,000 - 3,000 BC Holocene climatic optimum, or Atlantic in northern Europe (B-S)
- 6,200 BC 8.2 kiloyear event cold
- 5,000 - 4,100 BC Older Peron warm and wet, global sea levels were 2.5 to 4 metres (8 to 13 feet) higher than the twentieth-century average
- 3,900 BC 5.9 kiloyear event dry and cold, ends Neolithic Subpluvial in North Africa
- 3,200 - 2,900 Piora Oscillation, cold, perhaps not global. Wetter in Europe, drier elsewhere
- 2,200 BC 4.2 kiloyear event dry
- 1800-1500 BC Middle Bronze Age Cold Epoch
- 900 - 300 BC Iron Age Cold Epoch cold in North Atlantic
- 250 BC - 400 AD Roman Warm Period

Common Era/AD

- 250 BC - 400 AD Roman Warm Period
- Climate changes of 535-536 (535-536 AD), sudden cooling and failure of harvests, perhaps caused by volcanic dust
- 900 - 1300 Medieval warm period, wet in Europe, arid in North America
- Great Famine of 1315–1317

- Various dates between 1250 and 1550 or later are held to mark the start of the Little ice age, ending at equally varied dates mostly after 1850, or still in progress.
  - o 1460 - 1550 Spörer Minimum cold
  - o 1656-1715 Maunder Minimum low sunspot activity
  - o 1790 - 1830 Dalton Minimum low sunspot activity, cold
  - o Year Without a Summer (1816), caused by volcanic dust
- 1850 - present Retreat of glaciers since 1850, Instrumental temperature record
- present and recent past Global warming, perhaps to be named the Anthropocene period.

Polar Amplification

Polar amplification is the greater temperature increases in the Arctic compared to the earth as a whole as a result of the effect of feedbacks and other processes It is not observed in the Antarctic, largely because the Southern Ocean acts as a heat sink and the lack of seasonal snow cover. It is common to see it stated that "Climate models generally predict amplified warming in polar regions", e.g. Doran et al.. However, climate models predict amplified warming for the Arctic but *only modest warming for Antarctica.* Runaway climate change.

Runaway climate change describes a theoretical scenario in which the climate system passes a threshold or tipping point, after which internal positive feedback effects cause the climate to continue changing without further external forcings. The runaway climate change continues until it is overpowered by negative feedback effects which cause the climate system to restabilise at a new state.

Runaway terms are occasionally used in relation to climate change events in climatological literature. More generally, uses for these terms are found in the engineering journals, in books, and in the news media. *Runaway* terms are also used in the planetary sciences to describe the conditions that led to the current greenhouse state of Venus.

The phrase "runaway climate change" is used to describe a theory in which positive feedbacks result in rapid climate change. It is sometimes used in relation to climate change events in climatological literature. It is also used in the popular media and by environmentalists with reference to concerns about rapid global warming. Some astronomers use the similar expression runaway greenhouse effect to describe a situation where the climate deviates catastrophically and permanently from the original state - as happened on Venus.

Related Terms

- Tipping Level or tipping point - Climate forcing (greenhouse gas amount) reaches a point such that no additional forcing is required for large climate change and impacts
- Point of No Return - Climate system reaches a point with irreversible climate impacts (irreversible on a practical time scale) Example: disintegration of large ice sheet

Feedbacks

The core of the concept of runaway climate change is the idea of a large positive feedback within the climate system. When a change in global temperature causes an event to occur which itself changes global temperature, this is referred to as a feedback effect. If this effect acts in the same direction as the original temperature change, it is a destabilising positive feedback (e.g. warming causing more warming); and if in the opposite direction, it is a stabilising negative feedback (e.g. warming causing a cooling effect). If a sufficiently strong net positive feedback occurs, it is said that a climate tipping point has been passed and the temperature will continue to change until the changed conditions result in negative feedbacks that restabilise the climate.

An example of a negative feedback is that radiation leaving the Earth increases in proportion to the fourth power of temperature, in accordance with the Stefan-Boltzmann law. This feedback always operates and is proportional to the fourth power of temperature. Therefore, while it may be overridden by positive feedbacks for comparatively small temperature changes it will dominate for larger temperature changes. An example of a positive feedback is the ice-albedo feedback, in which increasing temperature causes ice to melt, which increases the amount of heat that Earth absorbs. This feedback only operates in a restricted range of temperatures (those for which ice exists, and does not cover the whole surface; once all the ice has melted, the feedback ceases to operate).

Climate feedback effects can be from:

- The same cause as the forcing (e.g. rising methane levels causing more methane to be released)
- Another greenhouse gas (e.g. CO_2 causing methane release)
- On other variables (e.g. ice-albedo feedback)

Without climate feedbacks, a doubling in atmospheric carbon dioxide (CO_2) concentration would result in a global average temperature increase of around 1.2°C. Water vapor amount and clouds are probably the most important global climate feedbacks. Historical information and global climate models indicate a climate sensitivity of 1.5 to 4.5°C, with a best estimate of 3°C. This is an amplification of the carbon dioxide forcing by a factor of 2.5. Some studies suggest a lower climate sensitivity, but other studies indicate a sensitivity above this range. Partly because of the difficulty in modelling the cloud feedback, the true climate sensitivity remains uncertain.

Examples

There are known examples of the Earth's climate producing a large response to small forcings; most obviously CO_2 feedback effect is believed to be part of the transition between glacial and interglacial periods, with the Milankovitch cycle providing the initial trigger. This is generally not considered to be a runaway climate change. Another example would be Dansgaard-Oeschger events.

Potentially unstable methane deposits exists in permafrost regions, which are expected to retreat as a result of global warming, and also clathrates, with the clathrate effect probably taking millennia to fully act. The potential role of methane from clathrates in near-future runaway scenarios is not certain, as studies show a slow release of methane, which may not be regarded as 'runaway' by all commentators. The clathrate gun runaway effect may be used to describe more rapid methane releases. Methane in the atmosphere has a high global warming potential, but breaks down relatively quickly to form CO_2, which is also a greenhouse gas. Therefore, slow methane release will have the long-term effect of adding CO_2 to the atmosphere.

In order to model clathrates and other reservoirs of greenhouse gases and their precursors, global climate models would have to be 'coupled' to a carbon cycle model. Some current global climate models do not include such modelling of methane deposits. A 2006 book chapter by Cox et al. considers the possibility of a future runaway climate feedback due to changes in the land carbon cycle:

Here we use a simple land carbon balance model to analyse the conditions required for a land sink-to-source transition, and address the question; could the land carbon cycle lead to a runaway climate feedback? [...] The simple land carbon balance model has effective parameters representing the sensitivities of climate and photosynthesis to CO_2, and the sensitivities of soil respiration and photosynthesis to temperature. This model is used to show that (a) a carbon sink-to-source transition is inevitable beyond some finite critical CO_2 concentration provided a few simple conditions are satisfied, (b) the value of the critical CO_2

concentration is poorly known due to uncertainties in land carbon cycle parameters and especially in the climate sensitivity to CO_2, and (c) that a true runaway land carbon-climate feedback (or linear instability) in the future is unlikely given that the land masses are currently acting as a carbon sink. Soil carbon and climate change: from the Jenkinson effect to the compost-bomb instability. More recently there has been a suggestion that the release of heat associated with soil decomposition, which is neglected in the vast majority of large-scale models, may be critically important under certain circumstances. Models with and without the extra self-heating from microbial respiration have been shown to yield significantly different results.

The present paper presents a mathematical analysis of a tipping point or runaway feedback that can arise when the heat from microbial respiration is generated more rapidly than it can escape from the soil to the atmosphere. This 'compost-bomb instability' is most likely to occur in drying organic soils with high porosity covered by an insulating lichen or moss layer. However, the instability is also found to be strongly dependent on the rate of global warming. This paper derives the conditions required to trigger the compost-bomb instability, and discusses the relevance of these to the concept of dangerous rates of climate change. On the basis of simple numerical experiments, rates of long-term warming equivalent to 10°C per century could be sufficient to trigger compost-bomb instability in drying organic soils.

Current Risk

The scientific consensus in the IPCC Fourth Assessment Report is that "Anthropogenic warming could lead to some effects that are abrupt or irreversible, depending upon the rate and magnitude of the climate change." Note however that this statement is about situations weaker than "runaway change".

Estimates of the size of the total carbon reservoir in Arctic permafrost and clathrates vary widely. It is suggested that at least 900 gigatonnes of carbon in permafrost exists worldwide. Furthermore, there are believed to be another 400 gigatonnes of carbon in methane clathrates in permafrost regions with 10,000 to 11,000 gigatonnes worldwide. This is large enough that if 10% of the stored methane were released, it would have an effect equivalent to a factor of 10 increase in atmospheric CO_2 concentrations. Methane is a potent greenhouse gas with a higher global warming potential than CO_2.

Worries about the release of this methane and carbon dioxide is linked to arctic shrinkage. Recent years have seen record low Arctic sea ice. It has been suggested that rapid melting of the sea ice may initiate a feedback loop that rapidly melts arctic permafrost. Methane clathrates on the sea-floor have also been predicted to destabilise, but much more slowly.

A release of methane from clathrates, however, is believed to be slow and chronic rather than catastrophic and that 21st-century effects of such a release are therefore likely to be 'significant but not catastrophic'. It is further noted that 'much methane from dissociated gas hydrate may never reach the atmosphere', as it can be dissolved into the ocean and be broken down biologically. Other research demonstrates that a release to the atmosphere can occur during large releases. These sources suggest that the clathrate gun effect alone will not be sufficient to cause 'catastrophic' climate change within a human lifetime. James E. Hansen has suggested that the Earth could experience a runaway greenhouse effect and adopt a climate like that of Venus if fossil-fuel use continues until reserves are exhausted.

Paleoclimatology

Events that could be described as runaway climate change may have occurred in the past.

Clathrate Gun

The clathrate gun hypothesis suggests an abrupt climate change due to a massive release of methane gas from methane clathrates on the seafloor. It has been speculated that the Permian-Triassic extinction event and the Paleocene-Eocene Thermal Maximum were caused by massive clathrate release.

Snowball Earth

Geological evidence shows that ice-albedo feedback caused sea ice advance to near the equator at several points in Earth history. Modelling work shows that such an event would indeed be a result of a self-sustaining ice-albedo effect, and that such a condition could be escaped via the accumulation of CO_2 from volcanic outgassing.

Climate Change Feedback

Climate change feedback is important in the understanding of global warming because feedback processes may amplify or diminish the effect of each climate forcing, and so play an important part in determining the overall climate sensitivity. Feedback in general is the process in which changing one quantity changes a second quantity, and the change in the second quantity in turn changes the first.

Positive feedback amplifies the change in the first quantity while negative feedback reduces it. By definition, forcings are external to the climate system while feedbacks are internal; in essence, feedbacks represent the internal processes of the system. Some feedbacks may act in relative isolation to the rest of the climate system; others may be tightly coupled; hence it may be difficult to tell just how much a particular process contributes. Forcings, feedbacks and the dynamics of the climate system determine how much and how fast the climate changes. The main positive feedback in global warming is the tendency of warming to increase the amount of water vapor in the atmosphere, which in turn leads to further warming. The main negative feedback comes from the Stefan–Boltzmann law, the amount of heat radiated from the Earth into space changes with the fourth power of the temperature of Earth's surface and atmosphere.

Some observed and potential effects of global warming are positive feedbacks, which contribute directly to further global warming. The Intergovernmental Panel on Climate Change's (IPCC) Fourth Assessment Report states that "Anthropogenic warming could lead to some effects that are abrupt or irreversible, depending upon the rate and magnitude of the climate change."

Carbon Cycle Feedbacks

There have been predictions, and some evidence, that global warming might cause loss of carbon from terrestrial ecosystems, leading to an increase of atmospheric CO_2 levels. Several climate models indicate that global warming through the 21st century could be accelerated by the response of the terrestrial carbon cycle to such warming. All 11 models in the C4MIP study found that a larger fraction of anthropogenic CO_2 will stay airborne if climate change is accounted for. By the end of the twenty-first century, this additional CO_2 varied between 20 and 200 ppm for the two extreme models, the majority of the models lying between 50 and 100 ppm.

The higher CO_2 levels led to an additional climate warming ranging between 0.1° and 1.5 °C. However, there was still a large uncertainty on the magnitude of these sensitivities. Eight models attributed most of the changes to the land, while three attributed it to the ocean. The strongest feedbacks in these cases are due to increased respiration of carbon from soils throughout the high latitude boreal forests of the Northern Hemisphere. One model in particular (HadCM3) indicates a secondary carbon cycle feedback due to the loss of much of the Amazon Rainforest in response to significantly reduced precipitation over tropical

South America. While models disagree on the strength of any terrestrial carbon cycle feedback, they each suggest any such feedback would accelerate global warming. Observations show that soils in England have been losing carbon at the rate of four million tonnes a year for the past 25 years according to a paper in Nature by Bellamy et al. in September 2005, who note that these results are unlikely to be explained by land use changes. Results such as this rely on a dense sampling network and thus are not available on a global scale. Extrapolating to all of the United Kingdom, they estimate annual losses of 13 million tons per year. This is as much as the annual reductions in carbon dioxide emissions achieved by the UK under the Kyoto Treaty (12.7 million tons of carbon per year).

It has also been suggested (by Chris Freeman) that the release of dissolved organic carbon (DOC) from peat bogs into water courses (from which it would in turn enter the atmosphere) constitutes a positive feedback for global warming. The carbon currently stored in peatlands (390–455 gigatonnes, one-third of the total land-based carbon store) is over half the amount of carbon already in the atmosphere. DOC levels in water courses are observably rising; Freeman's hypothesis is that, not elevated temperatures, but elevated levels of atmospheric CO_2 are responsible, through stimulation of primary productivity.

Tree deaths are believed to be increasing as a result of climate change, which is a positive feedback effect. This contradicts the previously widely held view that increased natural vegetation would lead to a negative-feedback effect.

Arctic Methane Release

Warming is also the triggering variable for the release of carbon (potentially as methane) in the arctic. Methane released from thawing permafrost such as the frozen peat bogs in Siberia, and from methane clathrate on the sea floor, creates a positive feedback.

Western Siberia is the world's largest peat bog, a one million square kilometre region of permafrost peat bog that was formed 11,000 years ago at the end of the last ice age. The melting of its permafrost is likely to lead to the release, over decades, of large quantities of methane. As much as 70,000 million tonnes of methane, an extremely effective greenhouse gas, might be released over the next few decades, creating an additional source of greenhouse gas emissions. Similar melting has been observed in eastern Siberia. Lawrence et al. (2008) suggest that a rapid melting of Arctic sea ice may start a feedback loop that rapidly melts Arctic permafrost, triggering further warming.

Methane Release From Hydrates

Methane clathrate, also called methane hydrate, is a form of water ice that contains a large amount of methane within its crystal structure. Extremely large deposits of methane clathrate have been found under sediments on the sea and ocean floors of Earth.

The sudden release of large amounts of natural gas from methane clathrate deposits, in a runaway global warming event, has been hypothesized as a cause of past and possibly future climate changes. The release of this trapped methane is a potential major outcome of a rise in temperature; it is thought that this might increase the global temperature by an additional 5° in itself, as methane is much more powerful as a greenhouse gas than carbon dioxide.

The theory also predicts this will greatly affect available oxygen content of the atmosphere. This theory has been proposed to explain the most severe mass extinction event on earth known as the Permian–Triassic extinction event, and also the Paleocene-Eocene Thermal Maximum climate change event. In 2008, a research expedition for the American Geophysical Union detected levels of methane up to 100 times above normal in the Siberian Arctic, likely being released by methane clathrates being released by holes in a frozen 'lid' of seabed permafrost, around the outfall of the Lena River and the area between the Laptev Sea and East Siberian Sea.

Abrupt Increases in Atmospheric Methane

Literature assessments by the Intergovernmental Panel on Climate Change (IPCC) and the US Climate Change Science Programme (CCSP) have considered the possibility of future projected climate change leading to a rapid increase in atmospheric methane. The IPCC Third Assessment Report, published in 2001, looked at possible rapid increases in methane due either to reductions in the atmospheric chemical sink or from the release of buried methane reservoirs.

In both cases, it was judged that such a release would be "exceptionally unlikely" (less than a 1% chance, based on expert judgement). The CCSP assessment, published in 2008, concluded that an abrupt release of methane into the atmosphere appeared "very unlikely" (less than 10% probability, based on expert judgement).

The CCSP assessment, however, noted that climate change would "very likely" (greater than 90% probability, based on expert judgement) accelerate the pace of persistent emissions from both hydrate sources and wetlands.

Decomposition

Organic matter stored in permafrost generates heat as it decomposes in response to the permafrost melting. This is significant mainly due to its effect on Arctic methane release.

Peat Decomposition

Peat, occurring naturally in peat bogs, is a store of carbon significant on a global scale. When peat dries it decomposes, and may additionally burn. Water table adjustment due to global warming may cause significant excursions of carbon from peat bogs. This may be released as methane, which can exacerbate the feedback effect, due to its high global warming potential.

Rainforest Drying

Rainforests, most notably tropical rainforests, are particularly vulnerable to global warming. There are a number of effects which may occur, but two are particularly concerning. Firstly, the drier vegetation may cause total collapse of the rainforest ecosystem. For example, the Amazon rainforest would tend to be replaced by caatinga ecosystems. Further, even tropical rainforests ecosystems which do not collapse entirely may lose significant proportions of their stored carbon as a result of drying, due to changes in vegetation.

Forest fires

The IPCC Fourth Assessment Report predicts that many mid-latitude regions, such as Mediterranean Europe, will experience decreased rainfall and an increased risk of drought, which in turn would allow forest fires to occur on larger scale, and more regularly. This releases more stored carbon into the atmosphere than the carbon cycle can naturally re-absorb, as well as reducing the overall forest area on the planet, creating a positive feedback loop. Part of that feedback loop is more rapid growth of replacement forests and a northward migration of forests as northern latitudes become more suitable climates for sustaining forests. There is a question of whether the burning of renewable fuels such as forests should be counted as contributing to global warming. Cook & Vizy also found that forest fires were likely in the Amazon Rainforest, eventually resulting in a transition to Caatinga vegetation in the Eastern Amazon region.

Desertification

Desertification is a consequence of global warming in some environments. Desert soils contain little humus, and support little vegetation. As a result, transition to desert ecosystems is typically associated with excursions of carbon.

CO_2 in the Oceans

Cooler water can absorb more CO_2 than warmer water. As ocean temperatures rise the oceans will absorb less CO_2 resulting in more warming. Conversely when cooler the oceans have absorbed more CO_2, resulting in further cooling. There is about 50 times more carbon in the oceans than there is in the atmosphere.

In addition to the water itself, the ecosystems of the oceans also sequester carbon. Their ability to do so is also expected to decline as the oceans warm: Warming reduces the nutrient levels of the mesopelagic zone (about 200 to 1000 m deep), which limits the growth of diatoms in favour of smaller phytoplankton that are poorer biological pumps of carbon.

Cloud Feedback

Warming is expected to change the distribution and type of clouds. Seen from below, clouds emit infrared radiation back to the surface, and so exert a warming effect; seen from above, clouds reflect sunlight and emit infrared radiation to space, and so exert a cooling effect. Whether the net effect is warming or cooling depends on details such as the type and altitude of the cloud. These details were poorly observed before the advent of satellite data and are difficult to represent in climate models.

Gas Release

Release of gases of biological origin may be affected by global warming, but research into such effects is at an early stage. Some of these gases, such as nitrous oxide released from peat, directly affect climate. Others, such as dimethyl sulfide released from oceans, have indirect effects.

Ice-albedo Feedback

When ice melts, land or open water takes its place. Both land and open water are on average less reflective than ice and thus absorb more solar radiation. This causes more warming, which in turn causes more melting, and this cycle continues. During times of global cooling, additional ice increases the reflectivity which reduces the absorption of solar radiation which results in more cooling in a continuing cycle. Considered a faster feedback mechanism.

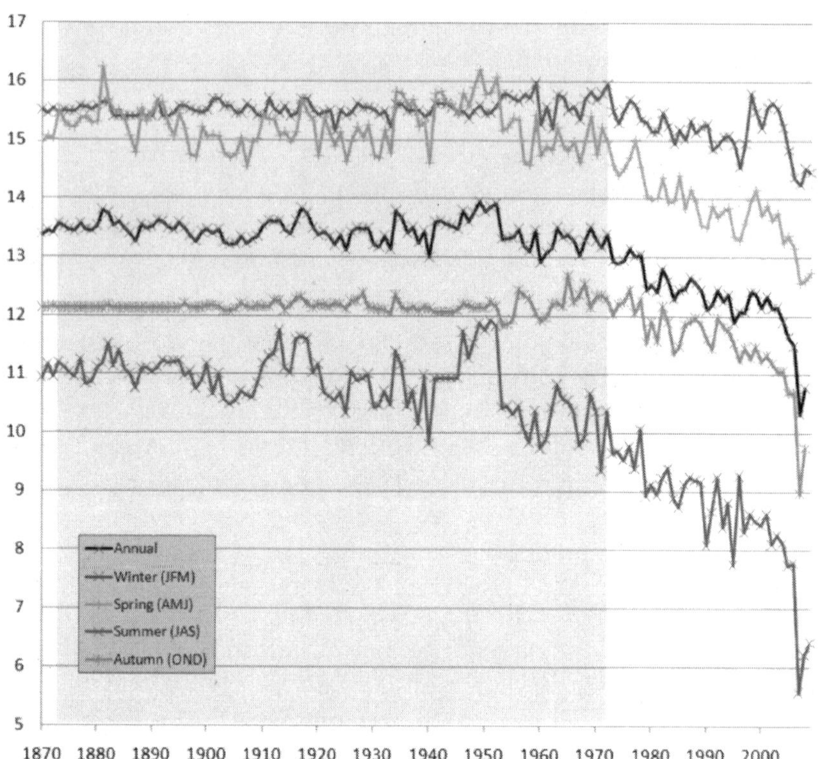

***Figure:** 1870-2009 Northern hemisphere sea ice extent in million square kilometres. Blue shading indicates the pre-satellite era; data then is less reliable. In particular, the near-constant level extent in Autumn up to 1940 reflects lack of data rather than a real lack of variation.*

Albedo change is also the main reason why IPCC predict polar temperatures in the northern hemisphere to rise up to twice as much as those of the rest of the world, in a process known as polar amplification. In September 2007, the Arctic sea ice area reached about half the size of the average summer minimum area between 1979 to 2000. Also in September 2007, Arctic sea ice retreated far enough for the Northwest Passage to become navigable to shipping for the first time in recorded history. The record losses of 2007 and 2008 may, however, be temporary.

Mark Serreze of the US National Snow and Ice Data Centre views 2030 as a "reasonable estimate" for when the summertime Arctic ice cap might be ice-free. The polar amplification of global warming is not predicted to occur in the southern hemisphere. The Antarctic sea ice reached its greatest extent on record since the beginning of observation in 1979, but the gain in ice in the south is exceeded by the loss in the north. The trend for global sea ice, northern hemisphere and southern hemisphere combined is clearly a decline.

Ice loss may have internal feedback processes, as melting of ice over land can cause eustatic sea level rise, potentially causing instability of ice shelves and inundating coastal ice masses, such as glacier tongues. Further, a potential feedback cycle exists due to earthquakes caused by isostatic rebound further destabilising ice shelves, glaciers and ice caps. The ice-albedo in some sub-arctic forests is also changing, as stands of larch (which shed their needles in winter, allowing sunlight to reflect off the snow in spring and fall) are being replaced by spruce trees (which retain their dark needles all year).

Water Vapor Feedback

If the atmospheres are warmed, the saturation vapor pressure increases, and the amount of water vapor in the atmosphere will tend to increase. Since water vapor is a greenhouse gas, the increase in water vapor content makes the atmosphere warm further; this warming causes the atmosphere to hold still more water vapor (a positive feedback), and so on until other processes stop the feedback loop. The result is a much larger greenhouse effect than that due to CO_2 alone. Although this feedback process causes an increase in the absolute moisture content of the air, the relative humidity stays nearly constant or even decreases slightly because the air is warmer. Climate models incorporate this feedback. Water vapor feedback is strongly positive, with most evidence supporting a magnitude of 1.5 to 2.0 $W/m^2/K$, sufficient to roughly double the warming that would otherwise occur. Considered a faster feedback mechanism.

Le Chatelier's Principle

Following Le Chatelier's principle, the chemical equilibrium of the Earth's carbon cycle will shift in response to anthropogenic CO_2 emissions. The primary driver of this is the ocean, which absorbs anthropogenic CO_2 via the so-called solubility pump. At present this accounts for only about one third of the current emissions, but ultimately most (~75%) of the CO_2 emitted by human activities will dissolve in the ocean over a period of centuries: "A better approximation of the lifetime of fossil fuel CO_2 for public discussion might be 300 years, plus 25% that lasts forever". However, the rate at which the ocean will take it up in the future is less certain, and will be affected by stratification induced by warming and, potentially, changes in the ocean's thermohaline circulation.

Chemical Weathering

Chemical weathering over the geological long term acts to remove CO_2 from the atmosphere. Biosequestration also captures and stores CO_2 by biological processes. The formation of shells by organisms in the ocean, over a very long time, removes CO_2 from the oceans. The complete conversion of CO_2 to limestone takes thousands to hundreds of thousands of years.

Net Primary Productivity

Net primary productivity changes in response to increased CO_2, as plants photosynthesis increased in response to increasing concentrations. However, this effect is swamped by other changes in the biosphere due to global warming.

Lapse Rate

The atmosphere's temperature decreases with height in the troposphere. Since emission of infrared radiation varies with temperature, longwave radiation escaping to space from the relatively cold upper atmosphere is less than that emitted towards the ground from the lower atmosphere. Thus, the strength of the greenhouse effect depends on the atmosphere's rate of temperature decrease with height. Both theory and climate models indicate that global warming will reduce the rate of temperature decrease with height, producing a negative *lapse rate feedback* that weakens the greenhouse effect. Measurements of the rate of temperature change with height are very sensitive to small errors in observations, making it difficult to establish whether the models agree with observations.

Blackbody Radiation

As the temperature of a black body increases, the emission of infrared radiation back into space increases with the fourth power of its absolute temperature according to Stefan–Boltzmann law. This increases the amount of outgoing radiation as the Earth warms. The impact of this negative feedback effect is included in global climate models summarized by the IPCC.

Runaway Greenhouse Effect

A runaway greenhouse effect is a process in which a net positive feedback between surface temperature and atmospheric opacity increases the strength of the greenhouse effect on a planet until its oceans boil away. An example of this is believed to have happened in the early history of Venus. On the Earth, the IPCC states that *a "runaway greenhouse effect" — analogous to Venus - appears to have virtually no chance of being induced by anthropogenic [human] activities.* Other, less catastrophic events, that nonetheless can produce a large climate change sometimes are loosely called a "runaway greenhouse" although it is not an appropriate description. For example, it has been hypothesized that large releases of greenhouse gases may have occurred concurrently with the Permian-Triassic extinction event or or Paleocene–Eocene Thermal Maximum. Other terms, such as 'abrupt climate change', or tipping points could be used when describing such scenarios.

Feedbacks

Positive feedbacks do not have to lead to a runaway effect, as the gain is not always sufficient. A strong negative feedback always exists (radiation from a planet increases in proportion to the fourth power of temperature, in accordance with the Stefan-Boltzmann law) so the positive feedback effect has to be very strong to cause a runaway effect. An increase in temperature from greenhouse gases leading to increased water vapor (which is itself a greenhouse gas) causing further warming is a positive feedback, but not a runaway effect, on Earth. Positive feedback effects are common (e.g. ice-albedo feedback) but runaway effects do not necessarily emerge from their presence.

Venus

A runaway greenhouse effect involving carbon dioxide and water vapor may have occurred on Venus. In this scenario, early Venus may have had a global ocean. As the brightness of the early Sun increased, the amount of water vapor in the atmosphere increased, increasing the temperature and consequently increasing the evaporation of the ocean, leading eventually to the situation in which the oceans boiled, and all of the water vapor entered the atmosphere. On Venus today there is little water vapor in the atmosphere. If water vapor did contribute to the warmth of Venus at one time, this water is thought to have escaped to space. Some evidence for this scenario comes from the extremely high Deuterium to Hydrogen ratio in Venus' atmosphere, roughly ~150x that of Earth, since light hydrogen would escape from the atmosphere more readily than its heavier isotope, Deuterium. Venus is sufficiently strongly heated by the Sun that water vapor can rise much higher in the atmosphere and be split into hydrogen and oxygen by ultraviolet light. The hydrogen can then escape from the atmosphere and the oxygen recombines. Carbon dioxide, the dominant greenhouse gas in the current Venusian atmosphere, owes its larger concentration to the weakness of carbon recycling as compared to Earth (which requires liquid water), where the carbon dioxide emitted from volcanoes is efficiently subducted into the Earth by plate tectonics on geologic time scales.

Earth

Earth's climate has swung repeatedly between warm periods and ice ages during its history. In the current climate the gain of the positive feedback effect from increased atmospheric water vapor is well below that which is required to boil away the oceans. Climate scientist John Houghton has written that "[there] is no possibility of [Venus's] runaway greenhouse conditions occurring on the Earth". However, climatologist James Hansen disagrees. In his *Storms of My Grandchildren* he says that burning coal and mining shale oil will result in runaway greenhouse on Earth. Benton and Twitchett's have a different definition of a runaway greenhouse; events meeting this definition have been suggested as a cause for the Paleocene-Eocene Thermal Maximum and the great dying.

Physics of the Runaway Greenhouse

Normally, when a planet's radiation balance is perturbed (e.g., by increasing the amount of sunlight it gets or changing the greenhouse concentration) it will transition to a new temperature until a stabilizing feedback known as the Stefan-Boltzmann response restores an equilibrium between the amount of energy the body absorbs and that which it emits. For example, if the Earth received more sunlight it would result in a temporary disequilibrium (more energy in than out) and result in warming. However, because the Stefan-Boltzmann response mandates that this hotter planet emit more energy, eventually a new radiation balance can be reached and the temperature will be maintained at its new, higher value.

However, when the planet has an operating water vapor feedback, the efficiency of the greenhouse effect increases as the temperature does, and so the outgoing radiation to space increases less rapidly than for a pure Stefan-Boltzmann radiator behaving like a blackbody. Eventually, the infrared absorption increases so much that the amount of energy escaping to space no longer depends on the temperature at the surface, and asymptotes to a fixed value. If the amount of energy that the planet receives from the star (or from internal heat sources) exceeds this value than radiative equilibrium can never be achieved and the result is a runaway, which continues until the water vapor feedback ceases (which may be when the entire ocean is evaporated).

Connection to Habitability

The concept of a habitable zone has been used by Planetary scientists and astrobiologists to define an orbital region around a star in which a planet (or moon) can sustain liquid water. Under this definition, the inner edge of the habitable zone (i.e., the closest point to a star that a planet can be until it can no longer sustains liquid water) is determined by the point in which the runaway greenhouse process occurs. For sun-like stars, this inner edge is estimated to reside at roughly 84% the distance from the Earth to the sun although feedbacks such as cloud-induced albedo increase could modify this estimate somewhat.

Homogenization (climate)

Homogenization in climate research means the removal of non-climatic changes. Next to changes in the climate itself, raw climate records also contain non-climatic jumps and changes for example due to relocations or changes in instrumentation. The most used principle to remove these inhomogeneities is the relative homogenization approach in which a candidate stations is compared to a reference time series based on one or more neighbouring stations. The candidate and reference station(s) experience about the same climate, non-climatic changes that happen only in one station can thus be identified and removed.

Climate Observations

To study climate change and variability long instrumental climate records are essential, but are best not used directly. These datasets are essential since they are the basis for assessing century-scale trends or for studying the natural (long-term) variability of climate, amongst others. The value of these datasets, however, strongly depends on the homogeneity of the underlying time series. A homogeneous climate record is one where variations are caused only by variations in weather and climate. Long instrumental records are rarely if ever homogeneous. Results from the homogenization of instrumental western climate records indicate that detected inhomogeneities in mean temperature series occur at a frequency of roughly 15 to 20 years. It should be kept in mind that most measurements have not been specifically made for climatic purposes, but rather to meet the needs of weather forecasting, agriculture and hydrology. Moreover the typical size of the breaks is often of the same order as the climatic change signal during the 20th century. Inhomogeneities are thus a significant source of uncertainty for the estimation of secular trends and decadal-scale variability. If all inhomogeneities would be purely random perturbations of the climate records, collectively their effect on the mean global climate signal would be negligible. However, certain changes are typical for certain periods and occurred in many stations, these are the most important causes as they can collectively lead to artificial biases in climate trends across large regions.

Causes of Inhomogeneities

The best known inhomogeneity is the urban heat island effect. The temperature in cities can be warmer than in the surrounding country side, especially at night. Thus as cities grow, one may expect that temperatures measured in cities become higher. On the other hand, with the advent of aviation, many meteorological offices and thus their stations have often been relocated from cities to nearby, typically cooler, airports.

Other non-climatic changes can be caused by changes in measurement methods. Meteorological instruments are typically installed in a screen to protect them from direct sun and wetting. In the 19th century it was common to use a metal screen in front of a window on a North facing wall. However, the building may warm the screen leading to higher temperature measurements. When this problem was realized the Stevenson screen was introduced, typically installed in gardens, away from buildings. This is still the most typical weather screen with its characteristic double-louvre door and walls for ventilation. The historical Montsouri and Wilds screens were used around 1900 and are open to the North and to the bottom. This improves ventilation, but it was found that infra-red radiation from the ground can influence the measurement on sunny calm days. Therefore, they are no longer used. Nowadays automatic weather stations, which reduce labour costs, are becoming more common; they protect the thermometer by a number of white plastic cones. This necessitated changes from manually recorded liquid and glass thermometers to automated electrical resistance thermometers, which reduced the recorded temperature values in the USA.

Also other climate elements suffer from inhomogeneities. The precipitation amounts observed in the early instrumental period, roughly before 1900, are biased and are 10% lower than nowadays because the precipitation measurements were often made on a roof. At the time, instruments were installed on rooftops to ensure that the instrument is never shielded from the rain, but it was found later that due to the turbulent flow of the wind on roofs, some rain droplets and especially snow flakes did not fall into the opening. Consequently measurements are nowadays performed closer to the ground.

Other typical causes of inhomogeneities are a change in measurement location; many observations, especially of precipitation are performed by volunteers in their garden or at their work place. Changes in the surrounding can often not be avoided, e.g., changes in the vegetation, the sealing of the land surface, and warm and sheltering buildings in the vicinity. There are also changes in measurement procedures such as the way the daily mean temperature is computed (by means of the minimum and maximum temperatures, or by averaging over 3 or 4 readings per day, or based on 10 minute data). Also changes in the observation times can lead to inhomogeneities. A recent review by Trewin focused on the causes of inhomogeneities.

The inhomogeneities are not always errors. This is seen most clear for stations affected by warming due to the urban heat island effect. From the perspective of global warming, such local effects are undesirable, but to study the influence of climate on health such measurements are fine. Other inhomogeneities are due to compromises that have to be made been ventilation and protection against the sun and wetting in the design of a weather shelter. Trying to reduce one type of error (for a certain weather condition) in the design will often lead to the more errors from the other factors. Meteorological measurements are not made in the laboratory. Small errors are inevitable and may not be relevant for meteorological purposes, but if such an error changes, it may well be an inhomogeneity for climatology.

Homogenization

To reliably study the real development of the climate, non-climatic changes have to be removed. The date of the change is often documented (called meta data: data about data), but not always. Meta data is often only available in the local language. In the best case, there are parallel measurements with the original and the new set-up for several years. This is a WMO (World Meteorological Organisation) guideline, but parallel measurements are unfortunately not very often performed, if only because the reason for stopping the original measurement is not known in advance, but probably more often to save money. By making parallel measurement with replicas of historical instruments, screens, etc. some of these inhomogeneities can still be studied today.

Because you are never sure that your meta data (station history) is complete, statistical homogenization should always be applied as well. The most commonly used statistical principle to detect and remove the effects of artificial changes is relative homogenization, which assumes that nearby stations are exposed to almost the same climate signal and that thus the differences between nearby stations can be utilized to detect inhomogeneities. By looking at the difference time series, the year to year variability of the climate is removed, as well as regional climatic trends. In such a difference time series, a clear and persistent jump of, for example 1°C, can easily be detected and can only be due to changes in the measurement conditions.

If there is a jump (break) in a difference time series, it is not yet clear which of the two stations it belongs to. Furthermore, time series typically have more than just one jump. These two features make statistical homogenization a challenging and beautiful statistical problem. Homogenization algorithms typically differ in how they try to solve these two fundamental problems.

In the past, it was customary to compute a composite reference time series computed from multiple nearby stations, compare this reference to the candidate series and assume that any jumps found are due to the candidate series. The latter assumption works because by using multiple stations as reference, the influence of inhomogeneities on the reference are much reduced. However, modern algorithms, no longer assume that the reference is homogeneous and can achieve better results this way. There are two main ways to do so. You can compute multiple composite reference time series from subsets of surrounding stations and test these references for homogeneity as well. Alternatively, you can only use pairs of stations and by comparing all pairs with each other determine which station most likely is the one with the break. If there is a break in 1950 in pair A&B and B&C, but not in A&C, the break is likely in station B; with more pairs such an inference can be made with more certainty.

If there are multiple breaks in a time series, the number of combinations easily becomes very large and it is becomes impossible to try them all. For example in case of five breaks (k=5) in 100 years of annual data (n=100), the number of combinations is about $100^5=10^{10}$ or 10 billion. This problem is sometimes solved iteratively/hierarchically, by first searching for the largest jump and then repeating the search in both sub-sections until they are too small. This does not always produce good results. A direct way to solve the problem is by an efficient optimization method called dynamic programming.

Sometimes there are no other stations in the same climate region. In this case, sometimes absolute homogenization is applied and the inhomogeneities are detected in the time series of one station. If there is a clear and large break at a certain date, you may well be able to correct it, but smaller jumps and gradually occurring inhomogeneities (urban heat island or a growing vegetation) cannot be distinguished from real natural variability and climate change. Data homogenized this way does not have the quality you may expect and should be used with much care.

Inhomogeneities in Climate Data

By homogenizing climate datasets, it was found that sometimes inhomogeneities can cause biased trends in raw data; that homogenization is indispensable to obtain reliable regional or global trends. For example, for the Greater Alpine Region a bias in the temperature trend between 1870s and 1980s of half a degree was found, which was due to *decreasing urbanization* of the network and systematic changes in the time of observation. The precipitation records of the early instrumental period are biased by -10% due to the systematic higher installation of the gauges at the time. Other possible bias sources are new types of weather shelters, the change from liquid and glass thermometers to electrical resistance thermometers, as well as the tendency to replace observers by automatic weather stations, the urban heat island effect and the transfer of many urban stations to airports.

In the project HOME homogenization algorithms were recently tested on artificial climate data with known inhomogeneities and it was found that relative homogenization improves climate data and that the modern methods that do not work with a homogeneous reference are most accurate.

Temperature Record of the Past 1000 Years

The temperature record of the 2nd millennium describes the reconstruction of temperatures since 1000 CE on the Northern Hemisphere, later extended back to 1 CE and also to cover the southern hemisphere. A reconstruction is needed because a reliable surface temperature record exists only since about 1850. Studying past climate is of interest for scientists in order to improve the understanding of current climate variability and, relatedly, providing a better basis for future climate projections. In particular, if the nature and magnitude of natural climate variability can be established, scientists will be able to better detect and attribute anthropogenic global warming. Note, however, that although temperature reconstructions from proxy data help us understand the character of natural climate variability, attribution of recent climate change relies on a broad range of methodologies of which the proxy reconstructions are only a small part. According to all major temperature reconstructions published in peer-reviewed journals, the increase in temperature in the 20th century and the temperature in the late 20th century is the highest in the record. Attention has tended to focus on the early work of Michael E. Mann, Bradley and Hughes (1998), whose "hockey stick" graph was featured in the 2001 United Nations Intergovernmental Panel on Climate Change report. The methodology and data sets used in creating the Mann *et al.* (1998) version of the hockey stick graph are disputed by Stephen McIntyre and Ross McKitrick, but the graph is overall acknowledged by the scientific community.

General Techniques and Accuracy

By far the best observed period is from 1850 to the present day, with coverage improving over time. Over this period the recent instrumental record, mainly based on direct thermometer readings, has approximately global coverage. It shows a general warming in global temperatures. Before this time various proxies must be used. These proxies are less accurate than direct thermometer measurements, have lower temporal resolution, and have less spatial coverage. Their only advantage is that they enable a longer record to be reconstructed. Since the direct temperature record is more accurate than the proxies (indeed, it is needed to calibrate them) it is used when available: i.e., from 1850 onwards.

Quantitative Methods Using Proxy Data

As there are few instrumental records before 1850, temperatures before then must be reconstructed based on proxy methods. One such method, based on principles of dendroclimatology, uses the width and other characteristics of tree rings to infer temperature. The isotopic composition of snow, corals, and stalactites can also be used to infer temperature. Other techniques which have been used include examining records of the time of crop harvests, the treeline in various locations, and other historical records to make inferences about the temperature. These proxy reconstructions are indirect inferences of temperature and thus tend to have greater uncertainty than instrumental data.

In general, the recent history of the proxy records is calibrated against local temperature records to estimate the relationship between temperature and the proxy. The longer history of the proxy is then used to reconstruct temperature from earlier periods. Proxy records must be averaged in some fashion if a global or hemispheric record is desired. Considerable care must be taken in the averaging process; for example, if a certain region has a large number of tree ring records, a simple average of all the data would strongly over-weight that region. Hence data-reduction techniques such as principal components analysis are used to combine some of these regional records before they are globally combined. An important distinction is between so-called 'multi-proxy' reconstructions, which attempt to obtain a global temperature reconstructions by using multiple proxy records distributed over the globe and more regional reconstructions. Usually, the various proxy records are combined arithmetically, in some weighted average. More recently, Osborn and Briffa used a simpler technique, counting the proportion of records that are positive, negative or neutral in any time period. This produces a result in general agreement with the conventional multi-proxy studies.

Several reconstructions suggest there was minimal variability in temperatures prior to the 20th century. More recently, Mann and Jones have extended their reconstructions to cover the 1st and 2nd millennia (GRL, 2003). The work was reproduced by Wahl and Ammann in 2005.

The Mann, Bradley and Hughes (1998) version of the temperature record is known as the "Hockey Stick" graph, first coined by Jerry Mahlman, director of the Geophysical Fluid Dynamics Laboratory. The work of Mann *et al.*, Jones *et al.*, Briffa and others forms a major part of the IPCC TAR's conclusion that "the rate and magnitude of global or hemispheric surface 20th century warming is likely to have been the largest of the millennium, with the 1990s and 1998 likely to have been the warmest decade and year". The IPCC AR4 concluced that "Average Northern Hemisphere temperatures during the second half of the 20th century were very likely higher than during any other 50-year period in the last 500 years and likely the highest in at least the past 1,300 years".

Qualitative Reconstruction Using Historical Records

It is also possible to use historical data such as times of grape harvests, sea-ice-free periods in harbours and diary entries of frost or heatwaves to produce indications of when it was warm or cold in particular regions. These records are harder to calibrate, are often only available sparsely through time, may be available only from developed regions, and are unlikely to come with good error estimates. These historical observations of the same time period show periods of both warming and cooling.

Astrophysicist Sallie Baliunas notes that these temperature variations correlate with solar variation and asserts that the number of observed sunspots give us a rough measure of how bright the sun is. Balunias and others have suggested that periods of decreased solar radiation are partially responsible for historically recorded periods of cooling such as the Maunder Minimum and the Little Ice Age. The same argument would imply that periods of increased solar radiation contributed to the Medieval Warm Period, when Greenland's icy coastal areas thawed enough to permit farming and colonisation.

Limitations

The apparent differences between the quantitative and qualitative approaches are not fully reconciled. The reconstructions mentioned above rely on various assumptions to generate their results. If these assumptions do not hold, the reconstructions would be unreliable. For quantitative reconstructions, the most fundamental assumptions are that proxy records vary with temperature and that non-temperature factors do not confound the results. In the historical records temperature fluctuations may be regional rather than hemispheric in scale.

In a letter to Nature (August 10, 2006) Bradley, Hughes and Mann pointed at the original title of their 1998 article: *Northern Hemisphere temperatures during the past millennium: inferences, uncertainties, and limitations* and pointed out more widespread high-resolution data are needed before more confident conclusions can be reached and that the uncertainties were the point of the article.

The Hockey Stick Controversy

There is an ongoing debate about the details of the temperature record and the means of its reconstruction, centered on the Mann, Bradley and Hughes (1998), "hockey stick" graph. Stephen McIntyre and Ross McKitrick claimed various errors in the methodology of Mann *et al.* (1998) and that the method of Mann, Bradley, and Hughes "when tested on persistent red noise, nearly always produces a hockey stick shaped first principal component". In turn, Michael E. Mann (supported by Tim Osborn, Keith Briffa and Phil Jones of the Climatic Research Unit) have disputed the claims made by McIntyre and McKitrick. The IPCC Fourth Assessment Report says that McIntyre and McKitrick "may have some theoretical foundation, but Wahl and Ammann (2006) also show that the impact on the amplitude of the final reconstruction is very small (~0.05°C)."

8

Global Cooling

Global cooling was a conjecture during the 1970s of imminent cooling of the Earth's surface and atmosphere along with a posited commencement of glaciation. This hypothesis had little support in the scientific community, but gained temporary popular attention due to a combination of a slight downward trend of temperatures from the 1940s to the early 1970s and press reports that did not accurately reflect the scientific understanding of ice age cycles. In contrast to the global cooling conjecture, the current scientific opinion on climate change is that the Earth has not durably cooled, but undergone global warming throughout the twentieth century.

Introduction: General Awareness and Concern

In the 1970s, there was increasing awareness that estimates of global temperatures showed cooling since 1945. Of those scientific papers considering climate trends over the 21st century, only 10% inclined towards future cooling, while most papers predicted future warming. The general public had little awareness of carbon dioxide's effects on climate, but Science News in May 1959 forecast a 25% increase in atmospheric carbon dioxide in the 150 years from 1850 to 2000, with a consequent warming trend. The actual increase in this period was 29%. Paul R. Ehrlich mentioned climate change from greenhouse gases in 1968. By the time the idea of global cooling reached the public press in the mid-1970s temperatures had stopped falling, and there was concern in the climatological community about carbon dioxide's warming effects. In response to such reports, the World Meteorological Organization issued a warning in June 1976 that *a very significant warming of global climate* was probable.

Currently there are some concerns about the possible *regional* cooling effects of a slowdown or shutdown of thermohaline circulation, which might be provoked by an increase of fresh water mixing into the North Atlantic due to glacial melting. The probability of this occurring is generally considered to be very low, and the IPCC notes, "even in models where the THC weakens, there is still a warming over Europe. For example, in all AOGCM integrations where the radiative forcing is increasing, the sign of the temperature change over north-west Europe is positive."

Physical Mechanisms

The cooling period is well reproduced by current (1999 on) global climate models (GCMs) that include the physical effects of sulphate aerosols, and there is now general agreement that aerosol effects were the dominant cause of the mid-20th century cooling. However, at the time there were two physical mechanisms that were most frequently advanced to cause cooling: aerosols and orbital forcing.

Aerosols

Human activity — mostly as a by-product of fossil fuel combustion, partly by land use changes — increases the number of tiny particles (aerosols) in the atmosphere. These have a *direct effect*: they effectively increase the planetary albedo, thus cooling the planet by reducing the solar radiation reaching the surface; and an *indirect effect*: they affect the properties of clouds by acting as cloud condensation nuclei. In the early 1970s some speculated that this cooling effect might dominate over the warming effect of the CO_2 release: see discussion of Rasool and Schneider (1971), below. As a result of observations and a switch to cleaner fuel burning, this no longer seems likely; current scientific work indicates that global warming is far more likely. Although the temperature drops foreseen by this mechanism have now been discarded in light of better theory and the observed warming, aerosols are believed to have contributed a cooling tendency (outweighted by increases in greenhouse gases) and also have contributed to "Global Dimming."

Orbital Forcing

Orbital forcing refers to the slow, cyclical changes in the tilt of Earth's axis and shape of its orbit. These cycles alter the total amount of sunlight reaching the earth by a small amount and affect the timing and intensity of the seasons. This mechanism is believed to be responsible for the timing of the ice age cycles, and understanding of the mechanism was increasing rapidly in the mid-1970s.

The seminal paper of Hays, Imbrie and Shackleton *Variations in the Earth's Orbit: Pacemaker of the Ice Ages* qualified its predictions with "forecasts must be qualified in two ways. First, they apply only to the natural component of future climatic trends - and not to anthropogenic effects such as those due to the burning of fossil fuels. Second, they describe only the long-term trends, because they are linked to orbital variations with periods of 20,000 years and longer. Climatic oscillations at higher frequencies are not predicted... the results indicate that the long-term trend over the next 20,000 years is towards extensive Northern Hemisphere glaciation and cooler climate".

The idea that ice ages cycles were predictable appears to have become conflated with the idea that another one was due "soon" - perhaps because much of this study was done by geologists, who are accustomed to dealing with very long time scales and use "soon" to refer to periods of thousands of years. A strict application of the Milankovitch theory does not allow the prediction of a "rapid" ice age onset (i.e., less than a century or two) since the fastest orbital period is about 20,000 years. Some creative ways around this were found, notably one championed by Nigel Calder under the name of "snowblitz", but these ideas did not gain wide acceptance.

It is common to see it asserted that the length of the current interglacial temperature peak is similar to the length of the preceding interglacial peak (Sangamon/Eem), and from this conclude that we might be nearing the end of this warm period. This conclusion is supported by the fact that the lengths of previous interglacials were regular; see appended figure. Petit et al. note that "Interglacials 5.5 and 9.3 are different from the Holocene, but similar to each other in duration, shape and amplitude." During each of these two events, there is a warm period of 4000 years followed by a relatively rapid cooling. *As an objection, the future orbital variations will not closely resemble those of the past.*

Concern in the 1960s and 1970s

Pre-1970s

J. Murray Mitchell showed as early as 1963 a multidecadal cooling since about 1940. At a conference on climate change held in Boulder, Colorado in 1965, evidence supporting Milankovitch cycles triggered speculation on how the calculated small changes in sunlight might somehow trigger ice ages. In 1966 Cesare Emiliani predicted that "a new glaciation will begin within a few thousand years."

In his 1968 book *The Population Bomb*, Paul R. Ehrlich wrote "The greenhouse effect is being enhanced now by the greatly increased level of carbon dioxide this is being countered by low-level clouds generated by contrails, dust, and other contaminants... At the moment we cannot predict what the overall climatic results will be of our using the atmosphere as a garbage dump."

1970s awareness

Concern peaked in the early 1970s, though "the possibility of anthropogenic warming dominated the peer-reviewed literature even then" (a cooling period began in 1945, and two decades of a cooling trend suggested a trough had been reached after several decades of warming). This peaking concern is partially attributable to the fact much less was then known about world climate and causes of ice ages. However, climate scientists were aware that predictions based on this trend were not possible - because the trend was poorly studied and not understood. Despite that, in the popular press the possibility of cooling was reported generally without the caveats present in the scientific reports, and "unusually severe winters in Asia and parts of North America in 1972 and 1973...pushed the issue into the public consciousness".

In the 1970s the compilation of records to produce hemispheric, or global, temperature records had just begun.

A history of the discovery of global warming states that: *While neither scientists nor the public could be sure in the 1970s whether the world was warming or cooling, people were increasingly inclined to believe that global climate was on the move, and in no small way.*

In 1972 Emiliani warned "Man's activity may either precipitate this new ice age or lead to substantial or even total melting of the ice caps..." By 1972 a group of glacial-epoch experts at a conference agreed that "the natural end of our warm epoch is undoubtedly near"; but the volume of Quaternary Research reporting on the meeting said that "the basic conclusion to be drawn from the discussions in this section is that the knowledge necessary for understanding the mechanism of climate change is still lamentably inadequate". Unless there were impacts from future human activity, they thought that serious cooling "must be expected within the next few millennia or even centuries"; but many other scientists doubted these conclusions. In 1972, George Kukla and Robert Matthews, in a Science write-up of a conference, asked when and how the current integlacial would end; concluding that "Global cooling and related rapid changes of environment, substantially exceeding the fluctuations experienced by man in historical times, must be expected within the next few millennia or even centuries."

1970 SCEP Report

The 1970 "Study of Critical Environmental Problems" reported the possibility of warming from increased carbon dioxide, but no concerns about cooling, setting a lower bound on the beginning of interest in "global cooling".

1971 to 1975: Papers on Warming and Cooling Factors

By 1971 studies indicated that human caused air pollution was spreading, but there was uncertainty as to whether aerosols would cause warming or cooling, and whether or not they were more significant than rising CO_2 levels. J. Murray Mitchell still viewed humans as "innocent bystanders" in the cooling from the 1940s to 1970, but in 1971 his calculations suggested that rising emissions could cause significant cooling after 2000, though he also argued that emissions could cause warming depending on circumstances. Calculations were too basic at this time to be trusted to give reliable results.

An early numerical computation of climate effects was published in the journal *Science* in July 1971 as a paper by S. Ichtiaque Rasool and Stephen H. Schneider, titled "Atmospheric Carbon Dioxide and Aerosols: Effects of Large Increases on Global Climate". The paper used rudimentary data and equations to compute the possible future effects of large increases in the densities in the atmosphere of two types of human environmental emissions:

1. greenhouse gases such as carbon dioxide;
2. particulate pollution such as smog, some of which remains suspended in the atmosphere in aerosol form for years.

The paper suggested that the global warming due to greenhouse gases would tend to have less effect with greater densities, and while aerosol pollution could cause warming, it was likely that it would tend to have a cooling effect which increased with density. They concluded that "An increase by only a factor of 4 in global aerosol background concentration may be sufficient to reduce the surface temperature by as much as 3.5 ° K. If sustained over a period of several years, such a temperature decrease over the whole globe is believed to be sufficient to trigger an ice age."

Both their equations and their data were badly flawed, as was soon pointed out by other scientists and confirmed by Schneider himself. In January 1972, Charlson et al. pointed out that with other reasonable assumptions, the model produced the opposite conclusion. The model made no allowance for changes in clouds or convection, and erroneously indicated that 8 times as much CO_2 would only cause 2°C of warming. In a paper published in 1975, Schneider corrected the overestimate of aerosol cooling by checking data on the effects of dust produced by volcanoes. When the model included estimated changes in solar intensity, it gave a reasonable match to temperatures over the previous thousand years and its prediction was that "CO_2 warming dominates the surface temperature patterns soon after 1980."

1972 and 1974 National Science Board

The National Science Board's *Patterns and Perspectives in Environmental Science* report of 1972 discussed the cyclical behaviour of climate, and the understanding at the time that the planet was entering a phase of cooling after a warm period. "Judging from the record of the past interglacial ages, the present time of high temperatures should be drawing to an end, to be followed by a long period of considerably colder temperatures leading into the next glacial age some 20,000 years from now." But it also continued; "However, it is possible, or even likely, that human interference has already altered the environment so much that the climatic pattern of the near future will follow a different path."

The Board's report of 1974, *Science And The Challenges Ahead,* continued on this theme. "During the last 20-30 years, world temperature has fallen, irregularly at first but more sharply over the last decade." However discussion of cyclic glacial periods does not feature in this report. Instead it is the role of man that is central to the report's analysis. "The cause of the cooling trend is not known with certainty. But there is increasing concern that man himself may be implicated, not only in the recent cooling trend but also in the warming temperatures over the last century". The report can not conclude whether carbon dioxide in warming, or agricultural and industrial pollution in cooling, are factors in the recent climatic changes, noting; "Before such questions as these can be resolved, major advances must be made in understanding the chemistry and physics of the atmosphere and oceans, and in measuring and tracing particulates through the system."

1975 National Academy of Sciences report

There also was a study by the U.S. National Academy of Sciences about issues that needed more research. This heightened interest in the fact that climate can change. The 1975 NAS report titled "Understanding

Climate Change: A Programme for Action" did not make predictions, stating in fact that "we do not have a good quantitative understanding of our climate machine and what determines its course. Without the fundamental understanding, it does not seem possible to predict climate." Its "programme for action" consisted simply of a call for further research, because "it is only through the use of adequately calibrated numerical models that we can hope to acquire the information necessary for a quantitative assessment of the climatic impacts."

The Report Further Stated

The climates of the earth have always been changing, and they will doubtless continue to do so in the future. How large these future changes will be, and where and how rapidly they will occur, we do not know.

The Science & Environmental Policy Project (SEPP), an organization which has no recognised scientific standing and has previously denied the link between tobacco and cancer, claims that "the NAS "experts" exhibited. .. hysterical fears" in the 1975 report.

1974 Time Magazine Article

While these discussions were ongoing in scientific circles, other accounts appeared in the popular media. In their June 24, 1974 issue, Time presented an article titled *Another Ice Age?* that noted "the atmosphere has been growing gradually cooler for the past three decades" but noted that "Some scientists... think that the cooling trend may be only temporary"

1975 Newsweek article

An April 28, 1975 article in *Newsweek* magazine was titled "The Cooling World", it pointed to "ominous signs that the Earth's weather patterns have begun to change" and pointed to "a drop of half a degree [Fahrenheit] in average ground temperatures in the Northern Hemisphere between 1945 and 1968." The article claimed "The evidence in support of these predictions [of global cooling] has now begun to accumulate so massively that meteorologists are hard-pressed to keep up with it." The *Newsweek* article did not state the cause of cooling; it stated that "what causes the onset of major and minor ice ages remains a mystery" and cited the NAS conclusion that "not only are the basic scientific questions largely unanswered, but in many cases we do not yet know enough to pose the key questions."

The article mentioned the alternative solutions of "melting the Arctic ice cap by covering it with black soot or diverting Arctic rivers" but conceded these were not feasible. The *Newsweek* article concluded by criticizing government leaders: "But the scientists see few signs that government leaders anywhere are even prepared to take the simple measures of stockpiling food or of introducing the variables of climatic uncertainty into economic projections of future food supplies...

The longer the planners (politicians) delay, the more difficult will they find it to cope with climatic change once the results become grim reality." The article emphasized sensational and largely unsourced consequences - "resulting famines could be catastrophic", "drought and desolation," "the most devastating outbreak of tornadoes ever recorded", "droughts, floods, extended dry spells, long freezes, delayed monsoons," "impossible for starving peoples to migrate," "the present decline has taken the planet about a sixth of the way towards the Ice Age."

On October 23, 2006, Newsweek issued a correction, over 31 years after the original article, stating that it had been "so spectacularly wrong about the near-term future" (though editor Jerry Adler claimed that 'the story wasn't "wrong" in the journalistic sense of "inaccurate."').

Other 1970s Sources

In the late 1970s there were several popular books on the topic, including *The Weather Conspiracy: The Coming of the New Ice Age.*

More Recent Climate Cooling Predictions

1980s

Concerns about nuclear winter arose in the early 1980s from several reports. Similar speculations have appeared over effects due to catastrophes such as asteroid impacts and massive volcanic eruptions. A prediction that massive oil well fires in Kuwait would cause significant effects on climate was quite incorrect.

1990s

The idea of a global cooling as the result of global warming was already proposed in the 1990s. In 2003, the Office of Net Assessment at the United States Department of Defence was commissioned to produce a study on the likely and potential effects of a modern climate change, especially of a shutdown of thermohaline circulation. The study, conducted under ONA head Andrew Marshall, modelled its prospective climate change on the 8.2 kiloyear event, precisely because it was the middle alternative between the Younger Dryas and the Little Ice Age. The study caused controversy in the media when it was made public in 2004. However, scientists acknowledge that "abrupt climate change initiated by Greenland ice sheet melting is not a realistic scenario for the 21st century".

Present Level of Knowledge

Currently, the concern that cooler temperatures would continue, and perhaps at a faster rate, has been observed to be incorrect by the IPCC. More has to be learned about climate, but the growing records have shown that the cooling concerns of 1975 have not been borne out. As for the prospects of the end of the current interglacial (again, valid only in the absence of human perturbations): it isn't true that interglacials have previously only lasted about 10,000 years; and Milankovitch-type calculations indicate that the present interglacial would probably continue for tens of thousands of years naturally. Other estimates (Loutre and Berger, based on orbital calculations) put the unperturbed length of the present interglacial at 50,000 years. Berger (EGU 2005 presentation) believes that the present CO_2 perturbation will last long enough to suppress the next glacial cycle entirely.

As the NAS report indicates, scientific knowledge regarding climate change was more uncertain than it is today. At the time that Rasool and Schneider wrote their 1971 paper, climatologists had not yet recognized the significance of greenhouse gases other than water vapor and carbon dioxide, such as methane, nitrous oxide, and chlorofluorocarbons. Early in that decade, carbon dioxide was the only widely studied human-influenced greenhouse gas. The attention drawn to atmospheric gases in the 1970s stimulated many discoveries in future decades. As the temperature pattern changed, global cooling was of waning interest by 1979.

Global Climate Model

A General Circulation Model (GCM) is a mathematical model of the general circulation of a planetary atmosphere or ocean and based on the Navier–Stokes equations on a rotating sphere with thermodynamic terms for various energy sources (radiation, latent heat). These equations are the basis for complex computer programmes commonly used for simulating the atmosphere or ocean of the Earth. Atmospheric and Oceanic GCMs (AGCM and OGCM) are key components of Global Climate Models along with sea ice and land-surface components. GCMs and global climate models are widely applied for weather forecasting, understanding the climate, and projecting climate change. Versions designed for decade to century time scale climate applications were originally created by Syukuro Manabe and Kirk Bryan at the Geophysical

Fluid Dynamics Laboratory in Princeton, New Jersey. These computationally intensive numerical models are based on the integration of a variety of fluid dynamical, chemical, and sometimes biological equations.

Note on Nomenclature

The acronym "GCM" can be used both for "Global Climate Model" and also for "General Circulation Model." While these do not refer to exactly the same thing, Global Circulation Models are typically the tools used for modelling climate, and hence the two terms are sometimes used as if they were interchangeable. For information on climate modelling,.

History: General Circulation Models

In 1956, Norman Phillips developed a mathematical model which could realistically depict monthly and seasonal patterns in the troposphere, which became the first successful climate model. Following Phillips's work, several groups began working to create general circulation models. The first general circulation climate model that combined both oceanic and atmospheric processes was developed in the late 1960s at the NOAA Geophysical Fluid Dynamics Laboratory. By the early 1980s, the United States' National Centre for Atmospheric Research had developed the Community Atmosphere Model; this model has been continuously refined into the 2000s.

In 1996, efforts began to initialize and model soil and vegetation types, which led to more realistic forecasts. Coupled ocean-atmosphere climate models such as the Hadley Centre for Climate Prediction and Research's HadCM3 model are currently being used as inputs for climate change studies. The importance of gravity waves was neglected within these models until the mid 1980s. Now, gravity waves are required within global climate models in order to properly simulate regional and global scale circulations, though their broad spectrum makes their incorporation complicated.

Atmospheric vs Oceanic Models

There are both atmospheric GCMs (AGCMs) and oceanic GCMs (OGCMs). An AGCM and an OGCM can be coupled together to form an atmosphere-ocean coupled general circulation model (CGCM or AOGCM). With the addition of other components (such as a sea ice model or a model for evapotranspiration over land), the AOGCM becomes the basis for a full climate model. Within this structure, different variations can exist, and their varying response to climate change may be studied (e.g., Sun and Hansen, 2003).

Modelling Trends

A recent trend in GCMs is to apply them as components of Earth System Models, e.g. by coupling to ice sheet models for the dynamics of the Greenland and Antarctic ice sheets, and one or more chemical transport models (CTMs) for species important to climate. Thus a carbon CTM may allow a GCM to better predict changes in carbon dioxide concentrations resulting from changes in anthropogenic emissions. In addition, this approach allows accounting for inter-system feedback: e.g. chemistry-climate models allow the possible effects of climate change on the recovery of the ozone hole to be studied. Climate prediction uncertainties depend on uncertainties in chemical, physical, and social models. Progress has been made in incorporating more realistic chemistry and physics in the models, but significant uncertainties and unknowns remain, especially regarding the future course of human population, industry, and technology. Note that many simpler levels of climate model exist; some are of only heuristic interest, while others continue to be scientifically relevant.

Model Structure

Three-dimensional (more properly four-dimensional) GCMs discretise the equations for fluid motion and integrate these forward in time. They also contain parameterisations for processes – such as convection – that occur on scales too small to be resolved directly. More sophisticated models may include representations of the carbon and other cycles.

A simple general circulation model (SGCM), a minimal GCM, consists of a dynamical core that relates material properties such as temperature to dynamical properties such as pressure and velocity. Examples are programmes that solve the primitive equations, given energy input into the model, and energy dissipation in the form of scale-dependent friction, so that atmospheric waves with the highest wavenumbers are the ones most strongly attenuated. Such models may be used to study atmospheric processes within a simplified framework but are not suitable for future climate projections.

Atmospheric GCMs (AGCMs) model the atmosphere (and typically contain a land-surface model as well) and impose sea surface temperatures (SSTs). A large amount of information including model documentation is available from AMIP. They may include atmospheric chemistry.

- AGCMs consist of a dynamical core which integrates the equations of fluid motion, typically for:

   - surface pressure
   - horizontal components of velocity in layers
   - temperature and water vapor in layers
- There is generally a radiation code, split into solar/short wave and terrestrial/infra-red/long wave
- Parametrizations are used to include the effects of various processes. All modern AGCMs include parameterizations for:
   - convection
   - land surface processes, albedo and hydrology
   - cloud cover

A GCM contains a number of prognostic equations that are stepped forward in time (typically winds, temperature, moisture, and surface pressure) together with a number of diagnostic equations that are evaluated from the simultaneous values of the variables. As an example, pressure at any height can be diagnosed by applying the hydrostatic equation to the predicted surface pressure and the predicted values of temperature between the surface and the height of interest. The pressure diagnosed in this way then is used to compute the pressure gradient force in the time-dependent equation for the winds.

Oceanic GCMs (OGCMs) model the ocean (with fluxes from the atmosphere imposed) and may or may not contain a sea ice model. For example, the standard resolution of HadOM3 is 1.25 degrees in latitude and longitude, with 20 vertical levels, leading to approximately 1,500,000 variables.

Coupled atmosphere-ocean GCMs (AOGCMs) (e.g. HadCM3, GFDL CM2.X) combine the two models. They thus have the advantage of removing the need to specify fluxes across the interface of the ocean surface. These models are the basis for sophisticated model predictions of future climate, such as are discussed by the IPCC. AOGCMs represent the pinnacle of complexity in climate models and internalise as many processes as possible. They are the only tools that could provide detailed regional predictions of future climate change. However, they are still under development. The simpler models are generally susceptible to simple analysis and their results are generally easy to understand. AOGCMs, by contrast, are often nearly as hard to analyse as the real climate system.

Model Grids

The fluid equations for AGCMs are discretised using either the finite difference method or the spectral method. For finite differences, a grid is imposed on the atmosphere. The simplest grid uses constant angular grid spacing (i.e., a latitude / longitude grid), however, more sophisticated non-rectantangular grids (e.g., icohedral) and grids of variable resolution are more often used. The "LMDz" model can be arranged to give high resolution over any given section of the planet. HadGEM1 (and other ocean models) use an ocean grid with higher resolution in the tropics to help resolve processes believed to be important for ENSO. Spectral models generally use a gaussian grid, because of the mathematics of transformation between spectral and grid-point space. Typical AGCM resolutions are between 1 and 5 degrees in latitude or longitude: the Hadley Centre model HadCM3, for example, uses 3.75 in longitude and 2.5 degrees in latitude, giving a grid of 96 by 73 points (96 x 72 for some variables); and has 19 levels in the vertical.

This results in approximately 500,000 "basic" variables, since each grid point has four variables (u,v, T, Q), though a full count would give more (clouds; soil levels). HadGEM1 uses a grid of 1.875 degrees in longitude and 1.25 in latitude in the atmosphere; HiGEM, a high-resolution variant, uses 1.25 x 0.83 degrees respectively. These resolutions are lower than is typically used for weather forecasting. Ocean resolutions tend to be higher, for example HadCM3 has 6 ocean grid points per atmospheric grid point in the horizontal. For a standard finite difference model, uniform gridlines converge towards the poles. This would lead to computational instabilities and so the model variables must be filtered along lines of latitude close to the poles. Ocean models suffer from this problem too, unless a rotated grid is used in which the North Pole is shifted onto a nearby landmass. Spectral models do not suffer from this problem. There are experiments using geodesic grids and icosahedral grids, which (being more uniform) do not have pole-problems. Another approach to solving the grid spacing problem is to deform a Cartesian cube such that it covers the surface of a sphere.

Flux Correction

Some early incarnations of AOGCMs required a somewhat *ad hoc* process of "flux correction" to achieve a stable climate (not all model groups used this technique). This resulted from separately prepared ocean and atmospheric models each having a different implicit flux from the other component than the other component could actually provide. If uncorrected this could lead to a dramatic drift away from observations in the coupled model. However, if the fluxes were 'corrected', the problems in the model that led to these unrealistic fluxes might be unrecognised and that might affect the model sensitivity. As a result, there has always been a strong disincentive to use flux corrections, and the vast majority of models used in the current round of the Intergovernmental Panel on Climate Change do not use them. The model improvements that now make flux corrections unnecessary are various, but include improved ocean physics, improved resolution in both atmosphere and ocean, and more physically consistent coupling between atmosphere and ocean models.

Convection

Moist convection causes the release of latent heat and is important to the Earth's energy budget. Convection occurs on too small a scale to be resolved by climate models, and hence must be parameterized. This has been done since the earliest days of climate modelling, in the 1950s. Akio Arakawa did much of the early work and variants of his scheme are still used although there are a variety of different schemes now in use. Clouds are typically parametrized, not because their physical processes are poorly understood, but because they occur on a scale smaller than the resolved scale of most GCMs. The causes and effects of their small scale actions on the large scale are represented by large scale parameters, hence "parameterization". The fact that cloud processes are not perfectly parameterized is due in part to a lack of perfect understanding of them, but parameterization itself is a fundamental and useful modelling technique, not a crutch. .

Output Variables

Most models include software to diagnose a wide range of variables for comparison with observations or study of atmospheric processes. An example is the 1.5 metre temperature, which is the standard height for near-surface observations of air temperature. This temperature is not directly predicted from the model but is deduced from the surface and lowest-model-layer temperatures. Other software is used for creating plots and animations.

Projections of Future Climate Change

Coupled ocean-atmosphere GCMs use transient climate simulations to project/predict future temperature changes under various scenarios. These can be idealised scenarios (most commonly, CO_2 increasing at 1%/yr) or more realistic (usually the "IS92a" or more recently the SRES scenarios). Which scenarios should be considered most realistic is currently uncertain, as the projections of future CO_2 (and sulphate) emission are themselves uncertain.

The 2001 IPCC Third Assessment Report figure shows the global mean response of 19 different coupled models to an idealised experiment in which CO_2 is increased at 1% per year. Figure shows the response of a smaller number of models to more realistic forcing. For the 7 climate models shown there, the temperature change to 2100 varies from 2 to 4.5 °C with a median of about 3 °C.

Future scenarios do not include unknowable events – for example, volcanic eruptions or changes in solar forcing. These effects are believed to be small in comparison to GHG forcing in the long term, but large volcanic eruptions, for example, are known to exert a temporary cooling effect. Human emissions of GHGs are an external input to the models, although it would be possible to couple in an economic model to provide these as well. Atmospheric GHG levels are usually supplied as an input, though it is possible to include a carbon cycle model including land vegetation and oceanic processes to calculate GHG levels.

Emissions Scenarios

For the six SRES marker scenarios, IPCC (2007:7–8) gave a "best estimate" of global mean temperature increase (2090–2099 relative to the period 1980–1999) that ranged from 1.8 °C to 4.0 °C. Over the same time period, the "likely" range (greater than 66% probability, based on expert judgement) for these scenarios was for a global mean temperature increase of between 1.1 and 6.4 °C.

Pope (2008) described a study where climate change projections were made using several different emission scenarios. In a scenario where global emissions start to decrease by 2010 and then decline at a sustained rate of 3% per year, the likely global average temperature increase was predicted to be 1.7 °C above pre-industrial levels by 2050, rising to around 2 °C by 2100. In a projection designed to simulate a future where no efforts are made to reduce global emissions, the likely rise in global average temperature was predicted to be 5.5 °C by 2100. A rise as high as 7 °C was thought possible but less likely. Sokolov *et al.* (2009) examined a scenario designed to simulate a future where there is no policy to reduce emissions. In their integrated model, this scenario resulted in a median warming over land (2090–2099 relative to the period 1980–1999) of 5.1 °C. Under the same emissions scenario but with different modelling of the future climate, the predicted median warming was 4.1 °C.

Accuracy of Models that Predict Global Warming

AOGCMs represent the pinnacle of complexity in climate models and internalise as many processes as possible. However, they are still under development and uncertainties remain. They may be coupled to models of other processes, such as the carbon cycle, so as to better model feedback effects. Most recent simulations show "plausible" agreement with the measured temperature anomalies over the past 150 years, when forced by observed changes in greenhouse gases and aerosols, but better agreement is achieved when natural forcings are also included. No model – whether a wind tunnel model for designing aircraft, or a climate model for projecting global warming – perfectly reproduces the system being modelled. Such inherently imperfect models may nevertheless produce useful results.

In this context, GCMs are capable of reproducing the general features of the observed global temperature over the past century. A debate over how to reconcile climate model predictions that upper air (tropospheric) warming should be greater than surface warming, with observations some of which appeared to show otherwise now appears to have been resolved in favour of the models, following revisions to the data.

The effects of clouds are a significant area of uncertainty in climate models. Clouds have *competing* effects on the climate. One of the roles that clouds play in climate is in cooling the surface by reflecting sunlight back into space; another is warming by increasing the amount of infrared radiation emitted from the atmosphere to the surface. In the 2001 IPCC report on climate change, the possible changes in cloud cover were highlighted as one of the dominant uncertainties in predicting future climate change.

Thousands of climate researchers around the world use climate models to understand the climate system. There are thousands of papers published about model-based studies in peer-reviewed journals – and a part of this research is work improving the models. Improvement has been difficult but steady (most obviously, state of the art AOGCMs no longer require flux correction), and progress has sometimes led to discovering new uncertainties.

In 2000, a comparison between measurements and dozens of GCM simulations of ENSO-driven tropical precipitation, water vapor, temperature, and outgoing longwave radiation found similarity between measurements and simulation of most factors. However the simulated change in precipitation was about one-fourth less than what was observed. Errors in simulated precipitation imply errors in other processes, such as errors in the evaporation rate that provides moisture to create precipitation.

The other possibility is that the satellite-based measurements are in error. Either indicates progress is required in order to monitor and predict such changes.

A more complete discussion of climate models is provided in the IPCC's Third Assessment Report.

- The model mean exhibits good agreement with observations.
- The individual models often exhibit worse agreement with observations.
- Many of the non-flux adjusted models suffered from unrealistic climate drift up to about 1 °C/century in global mean surface temperature.
- The errors in model-mean surface air temperature rarely exceed 1 °C over the oceans and 5 °C over the continents; precipitation and sea level pressure errors are relatively greater but the magnitudes and patterns of these quantities are recognisably similar to observations.
- Surface air temperature is particularly well simulated, with nearly all models closely matching the observed magnitude of variance and exhibiting a correlation > 0.95 with the observations.
- Simulated variance of sea level pressure and precipitation is within ±25% of observed.
- All models have shortcomings in their simulations of the present day climate of the stratosphere, which might limit the accuracy of predictions of future climate change.
  - o There is a tendency for the models to show a global mean cold bias at all levels.
  - o There is a large scatter in the tropical temperatures.
  - o The polar night jets in most models are inclined poleward with height, in noticeable contrast to an equatorward inclination of the observed jet.
  - o There is a differing degree of separation in the models between the winter sub-tropical jet and the polar night jet.
  - For nearly all models the r.m.s. error in zonal- and annual-mean surface air temperature is small compared with its natural variability.
  - o There are problems in simulating natural seasonal variability. (2000)

- In flux-adjusted models, seasonal variations are simulated to within 2 K of observed values over the oceans. The corresponding average over non-flux-adjusted models shows errors up to about 6 K in extensive ocean areas.
- Near-surface land temperature errors are substantial in the average over flux-adjusted models, which systematically underestimates (by about 5 K) temperature in areas of elevated terrain. The corresponding average over non-flux-adjusted models forms a similar error pattern (with somewhat increased amplitude) over land.
- In Southern Ocean mid-latitudes, the non-flux-adjusted models overestimate the magnitude of January-minus-July temperature differences by ~5 K due to an overestimate of summer (January) near-surface temperature. This error is common to five of the eight non-flux-adjusted models.
- Over Northern Hemisphere mid-latitude land areas, zonal mean differences between July and January temperatures simulated by the non-flux-adjusted models show a greater spread (positive and negative) about observed values than results from the flux-adjusted models.
- The ability of coupled GCMs to simulate a reasonable seasonal cycle is a necessary condition for confidence in their prediction of long-term climatic changes (such as global warming), but it is not a sufficient condition unless the seasonal cycle and long-term changes involve similar climatic processes.
- Coupled climate models do not simulate with reasonable accuracy clouds and some related hydrological processes (in particular those involving upper tropospheric humidity). Problems in the simulation of clouds and upper tropospheric humidity, remain worrisome because the associated processes account for most of the uncertainty in climate model simulations of anthropogenic change.

The precise magnitude of future changes in climate is still uncertain; for the end of the 21st century (2071 to 2100), for SRES scenario A2, the change of global average SAT change from AOGCMs compared with 1961 to 1990 is +3.0 °C (4.8 °F) and the range is +1.3 to +4.5 °C (+2 to +7.2 °F).

Relation to Weather Forecasting

The global climate models used for climate projections are very similar in structure to (and often share computer code with) numerical models for weather prediction but are nonetheless logically distinct. Most weather forecasting is done on the basis of interpreting the output of numerical model results. Since forecasts are short—typically a few days or a week—such models do not usually contain an ocean model but rely on imposed SSTs. They also require accurate initial conditions to begin the forecast—typically these are taken from the output of a previous forecast, with observations blended in. Because the results are needed quickly the predictions must be run in a few hours; but because they only need to cover a week of real time these predictions can be run at higher resolution than in climate mode.

Currently the ECMWF runs at 40 km (25 mi) resolution as opposed to the 100-to-200 km (62-to-120 mi) scale used by typical climate models. Often nested models are run forced by the global models for boundary conditions, to achieve higher local resolution: for example, the Met Office runs a mesoscale model with an 11 km (6.8 mi) resolution covering the UK, and various agencies in the U.S. also run nested models such as the NGM and NAM models. Like most global numerical weather prediction models such as the GFS, global climate models are often spectral models instead of grid models. Spectral models are often used for global models because some computations in modelling can be performed faster thus reducing the time needed to run the model simulation.

Computations Involved

Climate models use quantitative methods to simulate the interactions of the atmosphere, oceans, land surface, and ice. They are used for a variety of purposes from study of the dynamics of the climate system to projections of future climate. All climate models take account of incoming energy as short wave electromagnetic radiation, chiefly visible and short-wave (near) infrared, as well as outgoing energy as long wave (far) infrared electromagnetic radiation from the earth. Any imbalance results in a change in temperature.

The most talked-about models of recent years have been those relating temperature to emissions of carbon dioxide. These models project an upward trend in the surface temperature record, as well as a more rapid increase in temperature at higher altitudes.

Three (or more properly, four since time is also considered) dimensional GCM's discretise the equations for fluid motion and energy transfer and integrate these over time. They also contain parametrisations for processes—such as convection—that occur on scales too small to be resolved directly.

Atmospheric GCMs (AGCMs) model the atmosphere and impose sea surface temperatures as boundary conditions. Coupled atmosphere-ocean GCMs (AOGCMs, e.g. HadCM3, EdGCM, GFDL CM2.X, ARPEGE-Climat) combine the two models.

Models can range from relatively simple to quite complex:

- A simple radiant heat transfer model that treats the earth as a single point and averages outgoing energy
- this can be expanded vertically (radiative-convective models), or horizontally
- finally, (coupled) atmosphere–ocean–sea ice global climate models discretise and solve the full equations for mass and energy transfer and radiant exchange.

This is not a full list; for example "box models" can be written to treat flows across and within ocean basins. Furthermore, other types of modelling can be interlinked, such as land use, allowing researchers to predict the interaction between climate and ecosystems.

Simplified Models of Climate

Box Models

Box models are simplified versions of complex systems, reducing them to boxes (or reservoirs) linked by fluxes. The boxes are assumed to be mixed homogeneously. Within a given box, the concentration of any chemical species is therefore uniform. However, the abundance of a species within a given box may vary as a function of time due to the input to (or loss from) the box or due to the production, consumption or decay of this species within the box.

Simple box models, i.e. box model with a small number of boxes whose properties (e.g. their volume) do not change with time, are often useful to derive analytical formulas describing the dynamics and steady-state abundance of a species. More complex box models are usually solved using numerical techniques. Box models are used extensively to model environmental systems or ecosystems and in studies of ocean circulation and the carbon cycle.

Zero-dimensional Models

A very simple model of the radiative equilibrium of the Earth is:

$$(1-a)S\pi r^2 = 4\pi r^2 \epsilon\sigma T^4$$

where

- the left hand side represents the incoming energy from the Sun

- the right hand side represents the outgoing energy from the Earth, calculated from the Stefan-Boltzmann law assuming a constant radiative temperature, T, that is to be found,

and

- S is the solar constant – the incoming solar radiation per unit area—about 1367 W·m
- a is the Earth's average albedo, measured to be 0.3.
- r is Earth's radius—approximately 6.371×10m
- π is the mathematical constant (3.141...)
- σ is the Stefan-Boltzmann constant—approximately 5.67×10^{e8} J·K^{e4}·m^{e2}·s^{e1}
- ε is the effective emissivity of earth, about 0.612

The constant πr^2 can be factored out, giving

$$(1-a)S = 4\epsilon\sigma T^4$$

Solving for the temperature,

$$T = \sqrt[4]{\frac{(1-a)S}{4\epsilon\sigma}}$$

This yields an average earth temperature of 288 K (15 °C; 59 °F). This is because the above equation represents the effective *radiative* temperature of the Earth (including the clouds and atmosphere). The use of effective emissivity and albedo account for the greenhouse effect.

This very simple model is quite instructive. For example, it easily determines what the effect on average earth temperature of changes in solar constant or change of albedo or effective earth emissivity would be in the absence of feedback effects. Using the simple formula, the percent change of the average amount of each parameter, considered independently, to cause a one degree Celsius change in steady-state average earth temperature (*i.e.*, the *climate sensitivity*) is as follows:

- Solar constant 1.4%
- Albedo 3.3%
- Effective emissivity 1.4%

The average emissivity of the earth is readily estimated from available data. The emissivities of terrestrial surfaces are all in the range of 0.96 to 0.99 (except for some small desert areas which may be as low as 0.7). Clouds, however, which cover about half of the earth's surface, have an average emissivity of about 0.5 (which must be reduced by the fourth power of the ratio of cloud absolute temperature to average earth absolute temperature) and an average cloud temperature of about 258 K (–15 °C; 5 °F). Taking all this properly into account results in an effective earth emissivity of about 0.64 (earth average temperature 285 K (12 °C; 53 °F)).

This simple model readily determines the effect of changes in solar output or change of earth albedo or effective earth emissivity on average earth temperature.

It says nothing, however about what might cause these things to change, and does not incorporate feedback effects. Zero-dimensional models do not address the temperature distribution on the earth or the factors that move energy about the earth.

Radiative-convective Models

The zero-dimensional model above, using the solar constant and given average earth temperature, determines the effective earth emissivity of long wave radiation emitted to space. This can be refined in the vertical to a zero-dimensional radiative-convective model, which considers two processes of energy transport:

- upwelling and downwelling radiative transfer through atmospheric layers that both absorb and emit infrared radiation
- upward transport of heat by convection (especially important in the lower troposphere).

The radiative-convective models have advantages over the simple model: they can determine the effects of varying greenhouse gas concentrations on effective emissivity and therefore the surface temperature. But added parameters are needed to determine local emissivity and albedo and address the factors that move energy about the earth.

Links:

- "Effect of Ice-Albedo Feedback on Global Sensitivity in a One-Dimensional Radiative-Convective Climate Model"

Higher-dimensional Models

The zero-dimensional model may be expanded to consider the energy transported horizontally in the atmosphere. This kind of model may well be zonally averaged. This model has the advantage of allowing a rational dependence of local albedo and emissivity on temperature – the poles can be allowed to be icy and the equator warm – but the lack of true dynamics means that horizontal transports have to be specified.

EMICs (Earth-system models of intermediate complexity)

Depending on the nature of questions asked and the pertinent time scales, there are, on the one extreme, conceptual, more inductive models, and, on the other extreme, general circulation models operating at the highest spatial and temporal resolution currently feasible. Models of intermediate complexity bridge the gap. One example is the Climber-3 model. Its atmosphere is a 2.5-dimensional statistical-dynamical model with $7.5° \times 22.5°$ resolution and time step of 1/2 a day; the ocean is MOM-3 (Modular Ocean Model) with a $3.75° \times 3.75°$ grid and 24 vertical levels.

Climate Modellers

A climate modeller is a person who designs, develops, implements, tests, maintains or exploits climate models. There are three major types of institutions where a climate modeller may be found:

- In a national meteorological service. Most national weather services have at least a climatology section.
- In a university. Departments that may have climate modellers on staff include atmospheric sciences, meteorology, climatology, or geography, amongst others.
- In national or international research laboratories specialising in this field, such as the National Centre for Atmospheric Research (NCAR, in Boulder, Colorado, USA), the Geophysical Fluid Dynamics Laboratory (GFDL, in Princeton, New Jersey, USA), the Hadley Centre for Climate Prediction and Research (in Exeter, UK), the Max Planck Institute for Meteorology in Hamburg, Germany, or the Institut Pierre-Simon Laplace (IPSL in Paris, France). The World Climate Research Programme (WCRP), hosted by the World Meteorological Organization (WMO), coordinates research activities on climate modelling worldwide.

Global-warming Potential

Global-warming potential (GWP) is a relative measure of how much heat a greenhouse gas traps in the atmosphere. It compares the amount of heat trapped by a certain mass of the gas in question to the amount of heat trapped by a similar mass of carbon dioxide. A GWP is calculated over a specific time interval, commonly 20, 100 or 500 years. GWP is expressed as a factor of carbon dioxide (whose GWP is standardized to 1). For example, the 20 year GWP of methane is 72, which means that if the same mass of methane and carbon dioxide were introduced into the atmosphere, that methane will trap 72 times more heat than the carbon dioxide over the next 20 years. The substances subject to restrictions under the Kyoto protocol either are rapidly increasing their concentrations in Earth's atmosphere or have a large GWP.

The GWP depends on the following factors:

- the absorption of infrared radiation by a given species
- the spectral location of its absorbing wavelengths
- the atmospheric lifetime of the species

Thus, a high GWP correlates with a large infrared absorption and a long atmospheric lifetime. The dependence of GWP on the wavelength of absorption is more complicated. Even if a gas absorbs radiation efficiently at a certain wavelength, this may not affect its GWP much if the atmosphere already absorbs most radiation at that wavelength. A gas has the most effect if it absorbs in a "window" of wavelengths where the atmosphere is fairly transparent. The dependence of GWP as a function of wavelength has been found empirically and published as a graph. Because the GWP of a greenhouse gas depends directly on its infrared spectrum, the use of infrared spectroscopy to study greenhouse gases is centrally important in the effort to understand the impact of human activities on global climate change.

Calculating the global-warming Potential

Just as radiative forcing provides a simplified means of comparing the various factors that are believed to influence the climate system to one another, global-warming potentials (GWPs) are one type of simplified index based upon radiative properties that can be used to estimate the potential future impacts of emissions of different gases upon the climate system in a relative sense. GWP is based on a number of factors, including the radiative efficiency (infrared-absorbing ability) of each gas relative to that of carbon dioxide, as well as the decay rate of each gas (the amount removed from the atmosphere over a given number of years) relative to that of carbon dioxide.

The radiative forcing capacity (RF) is the amount of energy per unit area, per unit time, absorbed by the greenhouse gas, that would otherwise be lost to space. It can be expressed by the formula:

$$RF = \sum_{n=1}^{100} Abs_i * F_i \,/\, (pathlength * density)$$

where the subscript i represents an interval of 10 inverse centimeters. Abs_i represents the integrated infrared absorbance of the sample in that interval, and F_i represents the RF for that interval. The Intergovernmental Panel on Climate Change (IPCC) provides the generally accepted values for GWP, which changed slightly between 1996 and 2001. An exact definition of how GWP is calculated is to be found in the IPCC's 2001 Third Assessment Report. The GWP is defined as the ratio of the time-integrated radiative forcing from the instantaneous release of 1 kg of a trace substance relative to that of 1 kg of a reference gas:

$$GWP(x) = \frac{\int_0^{TH} a_x \cdot [x(t)]\,dt}{\int_0^{TH} a_r \cdot [r(t)]\,dt}$$

where TH is the time horizon over which the calculation is considered; a_x is the radiative efficiency due to a unit increase in atmospheric abundance of the substance (i.e., Wm^{d2} kg^{d1}) and [x(t)] is the time-dependent decay in abundance of the substance following an instantaneous release of it at time t=0. The denominator contains the corresponding quantities for the reference gas (i.e. CO_2). The radiative efficiencies a_x and a_r are not necessarily constant over time. While the absorption of infrared radiation by many greenhouse gases varies linearly with their abundance, a few important ones display non-linear behaviour for current and likely future abundances (e.g., CO_2, CH_4, and N_2O). For those gases, the relative radiative forcing will depend upon abundance and hence upon the future scenario adopted.

Since all GWP calculations are a comparison to CO_2 which is non-linear, all GWP values are affected. Assuming otherwise as is done above will lead to lower GWPs for other gases than a more detailed approach would.

Use in Kyoto Protocol

Under the Kyoto Protocol, the Conference of the Parties decided (decision 2/CP.3) that the values of GWP calculated for the IPCC Second Assessment Report are to be used for converting the various greenhouse gas emissions into comparable CO_2 equivalents when computing overall sources and sinks.

Importance of Time Horizon

Note that a substance's GWP depends on the timespan over which the potential is calculated. A gas which is quickly removed from the atmosphere may initially have a large effect but for longer time periods as it has been removed becomes less important. Thus methane has a potential of 25 over 100 years but 72 over 20 years; conversely sulfur hexafluoride has a GWP of 22,800 over 100 years but 16,300 over 20 years (IPCC TAR). The GWP value depends on how the gas concentration decays over time in the atmosphere. This is often not precisely known and hence the values should not be considered exact. For this reason when quoting a GWP it is important to give a reference to the calculation. The GWP for a mixture of gases can not be determined from the GWP of the constituent gases by any form of simple linear addition.

Commonly, a time horizon of 100 years is used by regulators (e.g., the California Air Resources Board).

Values

Carbon dioxide has a GWP of exactly 1 (since it is the baseline unit to which all other greenhouse gases are compared).

Although water vapour has a significant influence with regard to absorbing infrared radiation (which is the green house effect; see greenhouse gas), its GWP is not calculated. Its concentration in the atmosphere mainly depends on air temperature. There is no possibility to directly influence atmospheric water vapour concentration.

Climate (Chemical Processes)

Solar radiation delivers not only the energy to drive the global circulation of the oceans and the atmosphere, but it is also responsible for maintaining the temperature of the earth. The average temperature is determined by the amount of radiation which is absorbed, and this in turn is a function of the chemical composition of the atmosphere. The average temperature is 33 °C (OR 60 °F) greater than it would be because of the presence of the so-called greenhouse gases. These include water vapor, carbon dioxide, methane, ozone, and the halocarbons. Of these, water vapor has the largest influence, followed by carbon dioxide. Carbon dioxide levels in the atmosphere, as measured from gas bubbles trapped in ice cores, has increased since the start of the industrial revolution in the 19th century from about 280 to about 360 parts per million (by volume).

Direct measurements over the last 30 years show an average increase of 1.5 ppm/year. This increase, which contributes to global warming, has been ascribed to burning of fossil fuels and to a lesser extent to deforestation. However, incomplete understanding of the global carbon cycle limits the capability to evaluate the sources and sinks of carbon dioxide. We know that carbon is exchanged on decadal to centennial time scales between the terrestrial biosphere, the atmosphere, and the oceans. The exchange is much slower between the oceans and its sediments and sedimentary rocks. Fossil fuel burning, cement manufacture, and changing land-use are activities that release carbon (as carbon dioxide) to the atmosphere. The flux of carbon dioxide generated by these activities is small compared with mean natural gross fluxes, but it is sufficient to rapidly increase atmospheric levels by 29 % over the past 200 years. Because the net uptake of carbon dioxide occurs slowly (time-scale of centuries for the deep ocean), any additions will have a long-lasting effect on concentration in the atmosphere. For example, if carbon dioxide emissions were held constant at present day levels, atmospheric concentrations are estimated to continue to rise for at least two centuries.

Key processes in the carbon cycle include:

- Release of fossil fuel carbon dioxide to the atmosphere.
- The exchange of carbon dioxide between the atmosphere and ocean.
- Transport of carbon dioxide from the surface waters to the interior of the ocean and subsequent long-term storage in the bottom sediments of the deep ocean.

9

Solar Variation

Solar variation is the change in the amount of radiation emitted by the Sun and in its spectral distribution over years to millennia. These variations have periodic components, the main one being the approximately 11-year solar cycle (or sunspot cycle). The changes also have aperiodic fluctuations. In recent decades, solar activity has been measured by satellites, while before it was estimated using 'proxy' variables. Scientists studying climate change are interested in understanding the effects of variations in the total and spectral solar irradiance on Earth and its climate.

Variations in total solar irradiance were too small to detect with technology available before the satellite era, although the small fraction in ultra-violet light has recently been found to vary significantly more than previously thought over the course of a solar cycle. Total solar output is now measured to vary (over the last three 11-year sunspot cycles) by approximately 0.1%, or about 1.3 Watts per square meter (W/m^2) peak-to-trough from solar maximum to solar minimum during the 11-year sunspot cycle. The amount of solar radiation received at the outer surface of Earth's atmosphere averages 1366 W/m^2. There are no direct measurements of the longer-term variation, and interpretations of proxy measures of variations differ. The intensity of solar radiation reaching Earth has been relatively constant through the last 2000 years, with variations estimated at around 0.1-0.2%. Solar variation, together with volcanic activity are hypothesized to have contributed to climate change, for example during the Maunder Minimum. However, changes in solar brightness are too weak to explain recent climate change. History of study into solar variations

The longest recorded aspect of solar variations are changes in sunspots. The first record of sunspots dates to around 800 BC in China and the oldest surviving drawing of a sunspot dates to 1128. In 1610, astronomers began using the telescope to make observations of sunspots and their motions. Initial study was focused on their nature and behaviour. Although the physical aspects of sunspots were not identified until the 20th century, observations continued. Study was hampered during the 17th century due to the low number of sunspots during what is now recognized as an extended period of low solar activity, known as the Maunder Minimum. By the 19th century, there was a long enough record of sunspot numbers to infer periodic cycles in sunspot activity. In 1845, Princeton University professors Joseph Henry and Stephen Alexander observed the Sun with a thermopile and determined that sunspots emitted less radiation than surrounding areas of the Sun. The emission of higher than average amounts of radiation later were observed from the solar faculae.

Around 1900, researchers began to explore connections between solar variations and weather on Earth. Of particular note is the work of Charles Greeley Abbot. Abbot was assigned by the Smithsonian Astrophysical Observatory (SAO) to detect changes in the radiation of the Sun. His team had to begin by inventing instruments to measure solar radiation. Later, when Abbot was head of the SAO, it established a solar station at Calama, Chile to complement its data from Mount Wilson Observatory. He detected 27 harmonic periods within the 273-month Hale cycles, including 7, 13, and 39 month patterns. He looked for connections to weather by means such as matching opposing solar trends during a month to opposing temperature and precipitation trends in cities. With the advent of dendrochronology, scientists such as Waldo S. Glockattempted to connect variation in tree growth to periodic solar variations in the extant record and infer long-term secular variability in the solar constant from similar variations in millennial-scale chronologies.

Statistical studies that correlate weather and climate with solar activity have been popular for centuries, dating back at least to 1801, when William Herschel noted an apparent connection between wheat prices and sunspot records. They now often involve high-density global datasets compiled from surface networks and weather satellite observations and/or the forcing of climate models with synthetic or observed solar variability to investigate the detailed processes by which the effects of solar variations propagate through the Earth's climate system.

Solar Activity and Irradiance Measurement

Direct irradiance measurements have only been available during the last three cycles and are based on a composite of many different observing satellites. However, the correlation between irradiance measurements and other proxies of solar activity make it reasonable to estimate past solar activity. Most important among these proxies is the record of sunspot observations that has been recorded since ~1610. Since sunspots and associated faculae are directly responsible for small changes in the brightness of the sun, they are closely correlated to changes in solar output. Direct measurements of radio emissions from the Sun at 10.7 cm also provide a proxy of solar activity that can be measured from the ground since the Earth's atmosphere is transparent at this wavelength. Lastly, solar flares are a type of solar activity that can impact human life on Earth by affecting electrical systems, especially satellites. Flares usually occur in the presence of sunspots, and hence the two are correlated, but flares themselves make only tiny perturbations of the solar luminosity.

Recently it has been claimed that the total solar irradiance is varying in ways that are not duplicated by changes in sunspot observations or radio emissions, though Willson, DeWitte, and others have pointed out that these shifts in irradiance may be no more than the result of calibration problems in the measuring satellites. These speculations also admit the possibility that a small long-term trend might exist in solar irradiance.

Sunspots

Sunspots are relatively dark areas on the radiating 'surface' (photosphere) of the Sun where intense magnetic activity inhibits convection and cools the photosphere. Faculae are slightly brighter areas that form around sunspot groups as the flow of energy to the photosphere is re-established and both the normal flow and the sunspot-blocked energy elevate the radiating 'surface' temperature. Scientists have speculated on possible relationships between sunspots and solar luminosity since the historical sunspot area record began in the 17th century. Correlations are now known to exist with decreases in luminosity caused by sunspots (generally < - 0.3 %) and increases (generally < + 0.05 %) caused both by faculae that are associated with active regions as well as the magnetically active 'bright network'.

Modulation of the solar luminosity by magnetically active regions was confirmed by satellite measurements of total solar irradiance (TSI) by the ACRIM1 experiment on the Solar Maximum Mission (launched in 1980). The modulations were later confirmed in the results of the ERB experiment launched on the Nimbus 7 satellite in 1978, and satellite observation of solar irradiance continues today with ACRIM-3 and other satellite measurements. Sunspots in magnetically active regions are cooler and 'darker' than the average photosphere and cause temporary decreases in TSI of as much as 0.3 %. Faculae in magnetically active regions are hotter and 'brighter' than the average photosphere and cause temporary increases in TSI. The net effect during periods of enhanced solar magnetic activity is increased radiant output of the sun because faculae are larger and persist longer than sunspots.

Conversely, periods of lower solar magnetic activity and fewer sunspots (such as the Maunder Minimum) may correlate with times of lower terrestrial irradiance from the sun. There had been some suggestion that variations in the solar diameter might also cause significant variations in output. But recent work, mostly from the Michelson Doppler Imager instrument on SOHO, shows these changes to be small, about 0.001%, much less than the effect of magnetic activity changes (Dziembowski et al., 2001). Various studies have been made using sunspot number (for which records extend over hundreds of years) as a proxy for solar output (for which good records only extend for a few decades).

Also, ground instruments have been calibrated by comparison with high-altitude and orbital instruments. Researchers have combined present readings and factors to adjust historical data. Other proxy data — such as the abundance of cosmogenic isotopes — have been used to infer solar magnetic activity and thus likely brightness. Sunspot activity has been measured using the Wolf number for about 300 years. This index (also known as the Zürich number) uses both the number of sunspots and the number of groups of sunspots to compensate for variations in measurement. A 2003 study by Ilya Usoskin of the University of Oulu, Finland found that sunspots had been more frequent since the 1940s than in the previous 1150 years.

Sunspot numbers over the past 11,400 years have been reconstructed using dendrochronologically dated radiocarbon concentrations. The level of solar activity during the past 70 years is exceptional — the last period of similar magnitude occurred around 9,000 years ago (during the warm Boreal period). The Sun was at a similarly high level of magnetic activity for only ~10% of the past 11,400 years, and almost all of the earlier high-activity periods were shorter than the present episode.

Table: *Solar activity events and approximate dates*

Event	*Start*	*End*
Oort minimum	1040	1080
Medieval maximum	1100	1250
Wolf minimum	1280	1350
Spörer Minimum	1450	1550
Maunder Minimum	1645	1715
Dalton Minimum	1790	1820
Modern Maximum present	1900	

A list of historical Grand minima of solar activity includes also Grand minima ca. 690 AD, 360 BC, 770 BC, 1390 BC, 2860 BC, 3340 BC, 3500 BC, 3630 BC, 3940 BC, 4230 BC, 4330 BC, 5260 BC, 5460 BC, 5620 BC, 5710 BC, 5990 BC, 6220 BC, 6400 BC, 7040 BC, 7310 BC, 7520 BC, 8220 BC, 9170 BC.

Solar Cycles

The sun undergoes various quasi-periodic changes, the principal one referred to as the solar cycle has an 11-year quasi-period. Only the 11 and closely related 22 year cycles are clear in the observations.

- *11 years:* Most obvious is a gradual increase and more rapid decrease of the number of sunspots over a period ranging from 9 to 12 years, called the Schwabe cycle, named after Heinrich Schwabe. Differential rotation of the sun's convection zone (as a function of latitude) consolidates magnetic flux tubes, increases their magnetic field strength and makes them buoyant.

  As they rise through the solar atmosphere they partially block the convective flow of energy, cooling their region of the photosphere, causing 'sunspots'. The Sun's apparent surface, the photosphere, radiates more actively when there are more sunspots. Satellite monitoring of solar luminosity since 1980 has shown there is a direct relationship between the solar activity (sunspot) cycle and luminosity with a solar cycle peak-to-peak amplitude of about 0.1 %.

  Luminosity has also been found to decrease by as much as 0.3 % on a 10 day timescale when large groups of sunspots rotate across the Earth's view and increase by as much as 0.05 % for up to 6 months due to faculae associated with the large sunspot groups.

- *22 years:* Hale cycle, named after George Ellery Hale. The magnetic field of the Sun reverses during each Schwabe cycle, so the magnetic poles return to the same state after two reversals.

Hypothesized Cycles

Periodicity of solar activity with a periods longer than the sunspot cycle has been proposed. Some of these proposed longer cycles include:

- 87 years (70–100 years): *Gleissberg cycle*, named after Wolfgang Gleißberg, is thought to be an amplitude modulation of the 11-year Schwabe Cycle (Sonnett and Finney, 1990), Braun, *et al.*, (2005).
- 210 years: *Suess cycle* (a.k.a. –de Vries cycle"). Braun, *et al.*, (2005).
- 2,300 years: *Hallstatt cycle*
- 6000 years (Xapsos and Burke, 2009).

Other patterns have been detected:

- In carbon-14: 105, 131, 232, 385, 504, 805, 2,241 years (Damon and Sonnett, 1991).
- During the Upper Permian 240 million years ago, mineral layers created in the Castile Formation show cycles of 2,500 years.

The sensitivity of climate to cyclical variations in solar forcing will be higher for longer cycles due to the thermal inertia of the oceans, which acts to damp high frequencies.

Using a phenomenological approach, Scafetta and West (2005) found that the climate is 1.5 times as sensitive to 22 year cyclical forcing relative to 11 year cyclical forcing, and that the thermal inertia of the oceans induces a lag of approximately 2.2 (± 2) years in cyclic climate response in the temperature data.

Predictions Based on Patterns

- Perry and Hsu (2000) proposed a simple model based on emulating harmonics by multiplying the basic 11-year cycle by powers of 2, which produced results similar to Holocene behaviour. Extrapolation suggests a gradual cooling during the next few centuries with intermittent minor warmups and a return to near Little Ice Age conditions within the next 500 years. This cool period then may be followed approximately 1,500 years from now by a return to altithermal conditions similar to the previous Holocene Maximum.

- There is weak evidence for a quasi-periodic variation in the sunspot cycle amplitudes with a period of about 90 years ("Gleisberg cycle"). These characteristics indicate that the next solar cycle should have a maximum smoothed sunspot number of about 145±30 in 2010 while the following cycle should have a maximum of about 70±30 in 2023.
- Because carbon-14 cycles are quasi periodic, Damon and Sonett (1989) predict future climate:

Solar irradiance, or insolation, is the amount of sunlight which reaches the Earth. The equipment used might measure optical brightness, total radiation, or radiation in various frequencies. Historical estimates use various measurements and proxies.

Cycle length	*Cycle name*	*Last positivecarbon-14*	*Next "warming" anomaly*
232	—?—	AD 1922 (cool)	AD 2038
208	Suess	AD 1898 (cool)	AD 2210
88	Gleisberg	AD 1986 (cool)	AD 2030

Solar Irradiance of Earth and Its Surface

There are two common meanings:

- the radiation reaching the upper atmosphere
- the radiation reaching some point within the atmosphere, including the surface.

Various gases within the atmosphere absorb some solar radiation at different wavelengths, and clouds and dust also affect it. Measurements above the atmosphere are needed to determine variations in solar output, to avoid the confounding effects of changes within the atmosphere. There is some evidence that sunshine at the Earth's surface has been decreasing in the last 50 years possibly caused by increased atmospheric pollution, whilst over roughly the same timespan solar output has been nearly constant.

Milankovitch Cycle Variations

Some variations in insolation are not due to solar changes but rather due to the Earth moving closer or further from the Sun, or changes in the latitudinal distribution of radiation. These have caused variations of as much as 25% (locally; global average changes are much smaller) in solar insolation over long periods. The most recent event was an axial tilt of 24° during boreal summer at near the time of the *Holocene climatic optimum*.

Solar Interactions with Earth

Multiple factors have affected terrestrial climate change, including internal forcingsand human influences such as greenhouse gas emissions and land use change on top of any effects of solar variability. There are several hypotheses for how solar variations may affect Earth. Some variations, such as changes in the size of the Sun, are presently only of interest in the field of astronomy.

Changes in Total Irradiance

- Total solar irradiance changes slowly on decadal and longer timescales.
- The variation during recent solar magnetic activity cycles has been about 0.1% (peak-to-peak).
- Variations corresponding to solar changes with periods of 9–13, 18–25, and >100 years have been detected in sea-surface temperatures.
- In contrast to older reconstructions, most recent reconstructions of total solar irradiance point to an only small increase of only about 0.05 % to 0.1 % between Maunder Minimum and the present.
- Different composite reconstructions of total solar irradiance observations by satellites show different trends since 1980;.

Changes in Ultraviolet Irradiance

- Ultraviolet irradiance (EUV) varies by approximately 1.5 percent from solar maxima to minima, for 200 to 300 nm UV.
- Energy changes in the UV wavelengths involved in production and loss of ozone have atmospheric effects.
- The 30 hPa atmospheric pressure level has changed height in phase with solar activity during the last 4 solar cycles.
- UV irradiance increase causes higher ozone production, leading to stratospheric heating and to poleward displacements in the stratospheric and tropospheric wind systems.
- A proxy study estimates that UV has increased by 4.3% since the Maunder Minimum.

Changes in the Solar wind and the Sun's Magnetic Flux

- A more active solar wind and stronger magnetic field reduces the cosmic rays striking the Earth's atmosphere.

- Variations in the solar wind affect the size and intensity of the heliosphere, the volume larger than the Solar System filled with solar wind particles.
- Cosmogenic production of C, ^{10}Be and ^{36}Cl show changes tied to solar activity.
- Cosmic ray ionization in the upper atmosphere does change, but significant effects are not obvious.
- As the solar coronal-source magnetic flux doubled during the past century, the cosmic-ray flux has decreased by about 15%.
- The Sun's total magnetic flux rose by a factor of 1.41 from 1964–1996 and by a factor of 2.3 since 1901.

Effects on Clouds

Changes in ionization affect the abundance of aerosols that serve as the nuclei of condensation for cloud formation.

During periods of low solar activity (during solar minima), more cosmic rays reach Earth, potentially creating ultra-small aerosol particles which are precursors tocloud condensation nuclei.

Clouds formed from greater amounts of condensation nuclei are brighter, longer lived, and likely to produce less precipitation. It has been speculated that a change in cosmic rays could cause an increase in certain types of clouds, affecting Earth's albedo.

- Galactic cosmic rays have been hypothesized to affect formation of clouds through possible effects on production of cloud condensation nuclei.
- Several percent variation in cosmic rays and in tropospheric ionization occurs when the interplanetary magnetic field changes over the solar cycle, greater than the typically 0.1% variation in total solar irradiance meanwhile.
- Particularly at high latitudes where the shielding effect of Earth's magnetic field is less, some studies suggest cosmic ray variation may impact terrestrial low altitude cloud cover (unlike a lack of correlation with high altitude clouds), making such partially influenced by the solar-driven interplanetary magnetic field (as well as passage through the galactic arms over longer timeframes).

Other Effects due to Solar Variation

Interaction of solar particles, the solar magnetic field, and the Earth's magnetic field, cause variations in the particle and electromagnetic fields at the surface of the planet. Extreme solar events can affect electrical devices. Weakening of the Sun's magnetic field is believed to increase the number of interstellar cosmic rays which reach Earth's atmosphere, altering the types of particles reaching the surface.

Geomagnetic Effects

The Earth's polar aurorae are visual displays created by interactions between the solar wind, the solar magnetosphere, the Earth's magnetic field, and the Earth's atmosphere. Variations in any of these affect aurora displays. Solar coronal mass ejections, associated with high solar activity, will produce enhanced auroral activity, and visible aurorae at lower latitudes than usual. Sudden changes can cause the intense disturbances in the Earth's magnetic fields which are called geomagnetic storms.

Solar Proton Events

Energetic protons can reach Earth within 30 minutes of a major flare's peak. During such a solar proton event, Earth is showered in energetic solar particles (primarily protons) released from the flare site. Some of these particles spiral down Earth's magnetic field lines, penetrating the upper layers of our atmosphere where they produce additional ionization and may produce a significant increase in the radiation environment.

Galactic Cosmic Rays

An increase in solar activity (more sunspots) is accompanied by an increase in the "solar wind," which is an outflow of ionized particles, mostly protons and electrons, from the sun. The Earth's geomagnetic field, the solar wind, and the solar magnetic field deflect galactic cosmic rays (GCR). A decrease in solar activity increases the GCR penetration of the troposphere and stratosphere. GCR particles are the primary source of ionization in the troposphere above 1 km (below 1 km, radon is a dominant source of ionization in many areas).

Levels of GCRs have been indirectly recorded by their influence on the production of carbon-14 and beryllium-10. The Hallstatt solar cycle length of approximately 2300 years is reflected by climatic Dansgaard-Oeschger events. The 80–90 year solar Gleissberg cycles appear to vary in length depending upon the lengths of the concurrent 11 year solar cycles, and there also appear to be similar climate patterns occurring on this time scale.

Carbon-14 production

The production of carbon-14 (radiocarbon: ^{14}C) also is related to solar activity. Carbon-14 is produced in the upper atmosphere when cosmic ray bombardment of atmospheric nitrogen (^{14}N) induces the Nitrogen to undergo â+ decay, thus transforming into an unusual isotope of Carbon with an atomic weight of 14 rather than the more common 12. Because cosmic rays are partially excluded from the Solar System by the outward sweep of magnetic fields in the solar wind, increased solar activity results in a reduction of cosmic rays reaching the Earth's atmosphere and thus reduces ^{14}C production. Thus the cosmic ray intensity and carbon-14 production vary inversely to the general level of solar activity.

Therefore, the atmospheric ^{14}C concentration is *lower* during sunspot maxima and *higher* during sunspot minima. By measuring the captured ^{14}C in wood and counting tree rings, production of radiocarbon relative to recent wood can be measured and dated. A reconstruction of the past 10,000 years shows that the ^{14}C production was much higher during the mid-Holocene 7,000 years ago and decreased until 1,000 years ago. In addition to variations in solar activity, the long term trends in carbon-14 production are influenced by changes in the Earth's geomagnetic field and by changes in carbon cycling within the biosphere (particularly those associated with changes in the extent of vegetation since the last ice age).

Solar Variation and Climate

Both long-term and short-term variations in solar activity are hypothesized to affect global climate, but it has proven extremely challenging to directly quantify the link between solar variation and the earth's climate. The topic continues to be a subject of active study.

As discussed above, there are three suggested mechanisms by which solar variations may have an effect on climate:

- Solar irradiance changes directly affecting the climate ("Radiative forcing"). This is generally considered to be a minor effect, as the amplitudes of the variations in solar irradiance are much too small to have significant effect absent some amplification process.
- Variations in the ultraviolet component. The UV component varies by more than the total, so if UV were for some (as yet unknown) reason having a disproportionate effect, this might explain a larger solar signal in climate.
- Effects mediated by changes in cosmic rays (which are affected by the solar wind) such as changes in cloud cover.

Early research attempted to find a correlation between weather and sunspot activity, mostly without no success. Later research has concentrated more on correlating solar activity with global temperature. Crucial to the understanding of possible solar impact on terrestrial climate is accurate measurement of solar forcing. Unfortunately accurate measurement of incident solar radiation is only available since the satellite era, and even that is open to dispute: different groups find different values, due to different methods of cross-calibrating measurements taken by instruments with different spectral sensitivity. Scafetta and Willson found significant variations of solar luminosity between 1980 and 2000. But Lockwood and Frohlich find that solar forcing has declined since 1987.

Effect on Global Warming

The Intergovernmental Panel on Climate Change (IPCC) Third Assessment Report (TAR) concluded that the measured magnitude of recent solar variation is much smaller than the amplification effect due to greenhouse gases but acknowledges in the same report that there is a low level of scientific understanding with respect to solar variation..

Estimates of long-term solar irradiance changes have decreased since the TAR. However, empirical results of detectable tropospheric changes have strengthened the evidence for solar forcing of climate change. The most likely mechanism is considered to be some combination of direct forcing by changes in total solar irradiance, and indirect effects of ultraviolet (UV) radiation on the stratosphere. Least certain are indirect effects induced by galactic cosmic rays.

In 2002, Lean *et al.* stated that while "There is ... growing empirical evidence for the Sun's role in climate change on multiple time scales including the 11-year cycle", "changes in terrestrial proxies of solar activity can occur in the absence of long-term (i.e., secular) solar irradiance changes ... because the stochastic response increases with the cycle amplitude, not because there is an actual secular irradiance change." They conclude that because of this, "long-term climate change may appear to track the amplitude of the solar activity cycles," but that "Solar radiative forcing of climate is reduced by a factor of 5 when the background component is omitted from historical reconstructions of total solar irradiance. ..This suggests that general circulation model (GCM) simulations of twentieth century warming may overestimate the role of solar irradiance variability." More recently, a study and review of existing literature published in Nature in September 2006 suggests that the evidence is solidly on the side of solar brightness having relatively little effect on global climate, with little likelihood of significant shifts in solar output over long periods of time. Lockwood and Fröhlich,

2007, find that there "is considerable evidence for solar influence on the Earth's pre-industrial climate and the Sun may well have been a factor in post-industrial climate change in the first half of the last century," but that "over the past 20 years, all the trends in the Sun that could have had an influence on the Earth's climate have been in the opposite direction to that required to explain the observed rise in global mean temperatures."

A paper by Benestad and Schmidt concludes that "the most likely contribution from solar forcing a global warming is 7 ± 1% for the 20th century and is negligible for warming since 1980." This paper disagrees with the conclusions of a Scafetta and West study, who claim that solar variability has a significant effect on climate forcing. Based on correlations between specific climate and solar forcing reconstructions, they argue that a "realistic climate scenario is the one described by a large preindustrial secular variability (*e.g.*, the paleoclimate temperature reconstruction by Moberg et al.) with the total solar irradiance experiencing low secular variability (as the one shown by Wang et al.). Under this scenario, according to Scafetta and West, the Sun might have contributed 50% of the observed global warming since 1900. Stott *et al.* estimate that the residual effects of the prolonged high solar activity during the last 30 years account for between 16% and 36% of warming from 1950 to 1999.

Solar Variation Theory

A 1997 U.S. National Academy of Sciences study concluded that variations in total solar irradiance (TSI) were the most likely cause of significant climate change in the pre-industrial era, before significant human-generated carbon dioxide was put into the atmosphere.

A 2007 paper by Scafetta and West correlating solar proxy data and lower tropospheric temperature for the preindustrial era, before significant anthropogenic greenhouse forcing, suggested that TSI variations may have contributed to 50% of the global warming observed between 1900 and 2000 (although they conclude "our estimates about the solar effect on climate might be overestimated and should be considered as an upper limit.") This contrasts with the results from global circulation models that predict solar forcing of climate through direct radiative forcing is too small to explain a significant contribution. The relative significance of solar variability and other forcings of climate change during the industrial era is an area of ongoing research.

In 2000, Peter Stott and other researchers at the Hadley Centre in the United Kingdom published a paper in which they reported on the most comprehensive model simulations to date of the climate of the 20th century. Their study looked at both "natural forcing agents" (solar variations and volcanic emissions) as well as "anthropogenic forcing" (greenhouse gases and sulphate aerosols). They found that "solar effects may have contributed significantly to the warming in the first half of the century although this result is dependent on the reconstruction of total solar irradiance that is used. In the latter half of the century, we find that anthropogenic increases in greenhouses gases are largely responsible for the observed warming, balanced by some cooling due to anthropogenic sulphate aerosols, with no evidence for significant solar effects."

Stott's team found that combining all of these factors enabled them to closely simulate global temperature changes throughout the 20th century. They predicted that continued greenhouse gas emissions would cause additional future temperature increases "at a rate similar to that observed in recent decades". It should be noted that their solar forcing included "spectrally resolved changes in solar irradiance" but not indirect effects mediated through cosmic rays (discussed above and in the following section); these ideas are still being fleshed out. In addition, the study notes "uncertainties in historical forcing" — in other words, past natural forcing may still be having a delayed warming effect, most likely due to the oceans.

A graphical representation of the relationship between natural and anthropogenic factors contributing to climate change appears in "Climate Change 2001: The Scientific Basis", a report by the Intergovernmental Panel on Climate Change (IPCC).

Stott's 2003 work mentioned in the model section above largely revised his assessment, and found a significant solar contribution to recent warming, although still smaller (between 16 and 36%) than that of the greenhouse gases. Sami Solanki, the director of the Max Planck Institute for Solar System Research in Katlenburg-Lindau, Germany said:

> *The sun has been at its strongest over the past 60 years and may now be affecting global temperatures... the brighter sun and higher levels of so-called "greenhouse gases" both contributed to the change in the Earth's temperature, but it was impossible to say which had the greater impact.*

Nevertheless, Solanki agrees with the scientific consensus that the marked upswing in temperatures since about 1980 is attributable to human activity.

> *"Just how large this role [of solar variation] is, must still be investigated, since, according to our latest knowledge on the variations of the solar magnetic field, the significant increase in the Earth's temperature since 1980 is indeed to be ascribed to the greenhouse effect caused by carbon dioxide."*

Maunder Minimum

One historical long-term correlation between solar activity and climate change is the 1645–1715 Maunder minimum, a period of little or no sunspot activity which partially overlapped the "Little Ice Age" during which cold weather prevailed in Europe. The Little Ice Age encompassed roughly the 16th to the 19th centuries It is debated whether the low solar activity caused the cooling, or whether the cooling was caused by other factors.

The Spörer Minimum has also been identified with a significant cooling period between 1460 and 1550. Other indicators of low solar activity during this period are levels of the isotopes carbon-14 and beryllium-10. On the other hand, in a 2012 paper, Miller *et al.* link the Little Ice Age to an "unusual 50-year-long episode with four large sulfur-rich explosive eruptions," and notes "large changes in solar irradiance are not required."

Recent research had suggested that a new 90-year Maunder minimum would result in a reduction of global average temperatures of about 0.3°C, which would not be enough to offset the ongoing and forecasted average global temperature increase due to increased forcing from rising levels of carbon dioxide (generally referred to as global warming).

Correlations to Solar Cycle Length

In 1991, Eigil Friis-Christensen and Knud Lassen of the Danish Meteorological Institute in Copenhagen claimed to see a strong correlation of the length of the solar cycle with temperature changes throughout the northern hemisphere. Initially, they used sunspot and temperature measurements from 1861 to 1989, but later found that climate records dating back four centuries supported their findings. They reported that the relationship appeared to account for nearly 80 per cent of the measured temperature changes over this period. Although correlations often can be found, the mechanism behind these correlations is a matter of speculation. In a 2003 paper "Solar activity and terrestrial climate: an analysis of some purported correlations" Peter Laut demonstrates problems with some of these correlation analyses. Damon and Laut report in *Eos* that the apparent strong correlations displayed on these graphs have been obtained by incorrect handling of the physical data. The graphs are still widely referred to in the literature, and their misleading character has not yet been generally recognized.

Damon and Laut stated that when the graphs are corrected for filtering errors, the sensational agreement with the recent global warming, which drew worldwide attention, has totally disappeared.

On May 6, 2000, *New Scientist* magazine reported that Lassen and astrophysicist Peter Thejll had updated Friis-Christensen and Lassen's 1991 research (which originally only went to 1989) and found that while the solar cycle still accounts for about half the temperature rise since 1900, it fails to explain a rise of 0.4 °C since 1980. "The curves diverge after 1980," Thejll said, "and it's a startlingly large deviation. Something else is acting on the climate.... It has the fingerprints of the greenhouse effect." Likewise a 2005 review by Benestad found that the solar cycle length does not follow Earth's global mean surface temperature.

Solar Variation and Weather

There are some suggestions that there may also be regional climate impacts due to the solar activity. Measurements from NASA's Solar Radiation and Climate Experiment show that solar UV output is more variable than the total solar irradiance. Climate modelling suggests that low solar activity may result in, for example, colder winters in the US and southern Europe and warmer winters in Canada and northern Europe, with little change in globally-averaged temperature. More broadly, links have been suggested between solar cycles, global climate and events like El Nino. In other research, Daniel J. Hancock and Douglas N. Yarger found "statistically significant relationships between the double [~21 year] sunspot cycle and the 'January thaw' phenomenon along the East Coast and between the double sunspot cycle and 'drought' (June temperature and precipitation) in the Midwest."

Recent research at CERN's CLOUD facility examined links between cosmic rays and cloud condensation nuclei, demonstrating the effect of high-energy particulate radiation in nucleating aerosol particles which are precursors to cloud condensation nuclei. Dr. Jasper Kirby, a team leader at CLOUD, said, "At the moment, it [the experiment] actually says nothing about a possible cosmic-ray effect on clouds and climate, but it's a very important first step."

1983–1994 data from the International Satellite Cloud Climatology Project (ISCCP) showed that global low cloud formation was highly correlated with galactic cosmic ray (GCR) flux; subsequent to this period, the correlation breaks down. Changes of 3–4% in cloudiness and concurrent changes in cloud top temperatures have been correlated to the 11 and 22 yearsolar (sunspot) cycles, with increased GCR levels during "antiparallel" cycles. Global average cloud cover change has been found to be 1.5–2%. Several studies of GCR and cloud cover variations have found positive correlation at latitudes greater than 50° and negative correlation at lower latitudes. However, not all scientists accept this correlation as statistically significant, and some that do attribute it to other solar variability (*e.g.* UV or total irradiance variations) rather than directly to GCR changes. Difficulties in interpreting such correlations include the fact that many aspects of solar variability change at similar times, and some climate systems have delayed responses.

Historical Perspective

Physicist and historian Spencer R. Weart in The Discovery of Global Warming (2003) writes: The study of [sun spot] cycles was generally popular through the first half of the century. Governments had collected a lot of weather data to play with and inevitably people found correlations between sun spot cycles and select weather patterns. If rainfall in England didn't fit the cycle, maybe storminess in New England would. Respected scientists and enthusiastic amateurs insisted they had found patterns reliable enough to make predictions. Sooner or later though every prediction failed. An example was a highly credible forecast of a dry spell in Africa during the sunspot minimum of the early 1930s. When the period turned out to be wet, a meteorologist later recalled "the subject of sunspots and weather relationships fell into dispute, especially among British meteorologists who witnessed the discomfiture of some of their most respected superiors." Even in the 1960s he said, "For a young [climate] researcher to entertain any statement of sun-weather relationships was to brand oneself a crank."

Volcanism

Volcanic eruptions release gases and particulates into the atmosphere. Eruptions large enough to affect climate occur on average several times per century, and cause cooling (by partially blocking the transmission of solar radiation to the Earth's surface) for a period of a few years. The eruption of Mount Pinatubo in 1991, the second largest terrestrial eruption of the 20th century (after the 1912 eruption of Novarupta) affected the climate substantially. Global temperatures decreased by about 0.5 °C (0.9 °F). The eruption of Mount Tambora in 1815 caused the Year Without a Summer. Much larger eruptions, known as large igneous provinces, occur only a few times every hundred million years, but may cause global warming and mass extinctions. Volcanoes are also part of the extended carbon cycle. Over very long (geological) time periods, they release carbon dioxide from the Earth's crust and mantle, counteracting the uptake by sedimentary rocks and other geological carbon dioxide sinks. The US Geological Survey estimates are that volcanic emissions are at a much lower level than the effects of current human activities, which generate 100-300 times the amount of carbon dioxide emitted by volcanoes. A review of published studies indicates that annual volcanic emissions of carbon dioxide, including amounts released from mid-ocean ridges, volcanic arcs, and hot spot volcanoes, are only the equivalent of 3 to 5 days of human caused output. The annual amount put out by human activities may be greater than the amount released by supererruptions, the most recent of which was the Toba eruption in Indonesia 74,000 years ago.

Although volcanoes are technically part of the lithosphere, which itself is part of the climate system, the IPCC explicitly defines volcanism as an external forcing agent.

Plate Tectonics

Over the course of millions of years, the motion of tectonic plates reconfigures global land and ocean areas and generates topography. This can affect both global and local patterns of climate and atmosphere-ocean circulation.

The position of the continents determines the geometry of the oceans and therefore influences patterns of ocean circulation. The locations of the seas are important in controlling the transfer of heat and moisture across the globe, and therefore, in determining global climate. A recent example of tectonic control on ocean circulation is the formation of the Isthmus of Panama about 5 million years ago, which shut off direct mixing between the Atlantic and Pacific Oceans. This strongly affected the ocean dynamics of what is now the Gulf Stream and may have led to Northern Hemisphere ice cover. During the Carboniferous period, about 300 to 360 million years ago, plate tectonics may have triggered large-scale storage of carbon and increased glaciation. Geologic evidence points to a "megamonsoonal" circulation pattern during the time of the supercontinent Pangaea, and climate modelling suggests that the existence of the supercontinent was conducive to the establishment of monsoons.

The size of continents is also important. Because of the stabilizing effect of the oceans on temperature, yearly temperature variations are generally lower in coastal areas than they are inland. A larger supercontinent will therefore have more area in which climate is strongly seasonal than will several smaller continents or islands.

Human Influences

In the context of climate variation, anthropogenic factors are human activities which affect the climate. The scientific consensus on climate change is "that climate is changing and that these changes are in large part caused by human activities," and it "is largely irreversible."

> *"Science has made enormous inroads in understanding climate change and its causes, and is beginning to help develop a strong understanding of current and potential impacts that will affect people today and in coming decades. This understanding is crucial because it allows decision makers to place climate change in the context of other large challenges facing the nation and the world. There are still some uncertainties, and there always will be in understanding a complex system like Earth's climate. Nevertheless, there is a strong, credible body of evidence, based on multiple lines of research, documenting that climate is changing and that these changes are in large part caused by human activities. While much remains to be learned, the core phenomenon, scientific questions, and hypotheses have been examined thoroughly and have stood firm in the face of serious scientific debate and careful evaluation of alternative explanations."*
>
> — *United States National Research Council, Advancing the Science of Climate Change*

Of most concern in these anthropogenic factors is the increase in CO_2 levels due to emissions from fossil fuel combustion, followed by aerosols (particulate matter in the atmosphere) and cement manufacture. Other factors, including land use, ozone depletion, animal agriculture and deforestation, are also of concern in the roles they play - both separately and in conjunction with other factors - in affecting climate, microclimate, and measures of climate variables.

Physical Evidence for and Examples of Climatic Change

Evidence for climatic change is taken from a variety of sources that can be used to reconstruct past climates. Reasonably complete global records of surface temperature are available beginning from the mid-late 19th century. For earlier periods, most of the evidence is indirect—climatic changes are inferred from changes in proxies, indicators that reflect climate, such as vegetation, ice cores, dendrochronology, sea level change, and glacial geology.

Temperature Measurements and Proxies

The instrumental temperature record from surface stations was supplemented by radiosonde balloons, extensive atmospheric monitoring by the mid-20th century, and, from the 1970s on, with global satellite data as well. The $^{18}O/^{16}O$ ratio in calcite and ice core samples used to deduce ocean temperature in the distant past is an example of a temperature proxy method, as are other climate metrics noted in subsequent categories.

Historical and Archaeological Evidence

Climate change in the recent past may be detected by corresponding changes in settlement and agricultural patterns. Archaeological evidence, oral history and historical documentscan offer insights into past changes in the climate. Climate change effects have been linked to the collapse of various civilizations.

Glaciers

Glaciers are considered among the most sensitive indicators of climate change. Their size is determined by a mass balance between snow input and melt output. As temperatures warm, glaciers retreat unless snow precipitation increases to make up for the additional melt; the converse is also true.

Glaciers grow and shrink due both to natural variability and external forcings. Variability in temperature, precipitation, and englacial and subglacial hydrology can strongly determine the evolution of a glacier in a particular season. Therefore, one must average over a decadal or longer time-scale and/or over a many individual glaciers to smooth out the local short-term variability and obtain a glacier history that is related to climate. A world glacier inventory has been compiled since the 1970s, initially based mainly on aerial photographs and maps but now relying more on satellites. This compilation tracks more than 100,000 glaciers covering a total area of approximately 240,000 km^2, and preliminary estimates indicate that the remaining ice cover is around 445,000 km². The World Glacier Monitoring Service collects data annually on glacier retreat and glacier mass balance From this data, glaciers worldwide have been found to be shrinking significantly, with strong glacier retreats in the 1940s, stable or growing conditions during the 1920s and 1970s, and again retreating from the mid 1980s to present.

The most significant climate processes since the middle to late Pliocene (approximately 3 million years ago) are the glacial and interglacial cycles. The present interglacial period (the Holocene) has lasted about 11,700 years. Shaped by orbital variations, responses such as the rise and fall of continental ice sheets and significant sea-level changes helped create the climate. Other changes, including Heinrich events, Dansgaard–Oeschger events and the Younger Dryas, however, illustrate how glacial variations may also influence climate without the orbital forcing.

Glaciers leave behind moraines that contain a wealth of material—including organic matter, quartz, and potassium that may be dated—recording the periods in which a glacier advanced and retreated. Similarly, by tephrochronological techniques, the lack of glacier cover can be identified by the presence of soil or volcanic tephra horizons whose date of deposit may also be ascertained.

Arctic Sea Ice Loss

The decline in Arctic sea ice, both in extent and thickness, over the last several decades is further evidence for rapid climate change. Sea ice is frozen seawater that floats on the ocean surface. It covers millions of square miles in the polar regions, varying with the seasons. In the Arctic, some sea ice remains year after year, whereas almost all Southern Ocean or Antarctic sea ice melts away and reforms annually. Satellite observations show that Arctic sea ice is now declining at a rate of 11.5 percent per decade, relative to the 1979 to 2000 average.

Vegetation

A change in the type, distribution and coverage of vegetation may occur given a change in the climate. Some changes in climate may result in increased precipitation and warmth, resulting in improved plant growth and the subsequent sequestration of airborne CO_2. A gradual increase in warmth in a region will lead to earlier flowering and fruiting times, driving a change in the timing of life cycles of dependent organisms. Conversely, cold will cause plant bio-cycles to lag. Larger, faster or more radical changes, however, may result in vegetation stress, rapid plant loss and desertification in certain circumstances. An example of this occurred during the Carboniferous Rainforest Collapse (CRC), an extinction event 300 million years ago. At this time vast rainforests covered the equatorial region of Europe and America. Climate change devastated these tropical rainforests, abruptly fragmenting the habitat into isolated 'islands' and causing the extinction of many plant and animal species. Satellite data available in recent decades indicates that global terrestrial net primary production increased by 6% from 1982 to 1999, with the largest portion of that increase in tropical ecosystems, then decreased by 1% from 2000 to 2009.

Pollen Analysis

Palynology is the study of contemporary and fossil palynomorphs, including pollen. Palynology is used to infer the geographical distribution of plant species, which vary under different climate conditions. Different groups of plants have pollen with distinctive shapes and surface textures, and since the outer surface of pollen is composed of a very resilient material, they resist decay. Changes in the type of pollen found in different layers of sediment in lakes, bogs, or river deltas indicate changes in plant communities. These changes are often a sign of a changing climate. As an example, palynological studies have been used to track changing vegetation patterns throughout the Quaternary glaciations and especially since the last glacial maximum.

Precipitation

Past precipitation can be estimated in the modern era with the global network of precipitation gauges. Surface coverage over oceans and remote areas is relatively sparse, but, reducing reliance on interpolation, satellite data has been available since the 1970s. Quantification of climatological variation of precipitation in prior centuries and epochs is less complete but approximated using proxies such as marine sediments, ice cores, cave stalagmites, and tree rings. Climatological temperatures substantially affect precipitation. For instance, during the Last Glacial Maximum of 18,000 years ago, thermal-driven evaporation from the oceans onto continental landmasses was low, causing large areas of extreme desert, including polar deserts (cold but with low rates of precipitation). In contrast, the world's climate was wetter than today near the start of the warm Atlantic Period of 8000 years ago.

Estimated global land precipitation increased by approximately 2% over the course of the 20th century, though the calculated trend varies if different time endpoints are chosen, complicated by ENSO and other oscillations, including greater global land precipitation in the 1950s and 1970s than the later 1980s and 1990s despite the positive trend over the century overall. Similar slight overall increase in global river runoff and in average soil moisture has been perceived.

Dendroclimatology

Dendroclimatology is the analysis of tree ring growth patterns to determine past climate variations. Wide and thick rings indicate a fertile, well-watered growing period, whilst thin, narrow rings indicate a time of lower rainfall and less-than-ideal growing conditions.

Ice Cores

Analysis of ice in a core drilled from a ice sheet such as the Antarctic ice sheet, can be used to show a link between temperature and global sea level variations. The air trapped in bubbles in the ice can also reveal the CO_2 variations of the atmosphere from the distant past, well before modern environmental influences. The study of these ice cores has been a significant indicator of the changes in CO_2 over many millennia, and continues to provide valuable information about the differences between ancient and modern atmospheric conditions.

Animals

Remains of beetles are common in freshwater and land sediments. Different species of beetles tend to be found under different climatic conditions. Given the extensive lineage of beetles whose genetic makeup has not altered significantly over the millennia, knowledge of the present climatic range of the different species, and the age of the sediments in which remains are found, past climatic conditions may be inferred.

Similarly, the historical abundance of various fish species has been found to have a substantial relationships with observed climatic conditions. Changes in the primary productivity of autotrophs in the oceans can affect marine food webs.

Sea Level Change

Global sea level change for much of the last century has generally been estimated using tide gauge measurements collated over long periods of time to give a long-term average. More recently, altimeter measurements — in combination with accurately determined satellite orbits — have provided an improved measurement of global sea level change. To measure sea levels prior to instrumental measurements, scientists have dated coral reefs that grow near the surface of the ocean, coastal sediments, marine terraces, ooids in limestones, and nearshore archaeological remains. The predominant dating methods used are uranium series and radiocarbon, with cosmogenic radionuclides being sometimes used to date terraces that have experienced relative sea level fall.

Climate Change Mitigation

Climate change mitigation is action to decrease the intensity of radiative forcing in order to reduce the potential effects of global warming. Mitigation is distinguished from adaptation to global warming, which involves acting to tolerate the effects of global warming. Most often, climate change mitigation scenarios involve reductions in the concentrations of greenhouse gases, either by reducing their sources or by increasing their sinks.

The UN defines mitigation in the context of climate change, as a human intervention to reduce the sources or enhance the sinks of greenhouse gases. Examples include using fossil fuels more efficiently for industrial processes or electricity generation, switching to renewable energy (solar energy or wind power), improving the insulation of buildings, and expanding forests and other "sinks" to remove greater amounts of carbon dioxide from the atmosphere.

Scientific consensus on global warming, together with the precautionary principle and the fear of abrupt climate change is leading to increased effort to develop new technologies and sciences and carefully manage others in an attempt to mitigate global warming. Most means of mitigation appear effective only for preventing further warming, not at reversing existing warming. The Stern Review identifies several ways of mitigating climate change. These include reducing demand for emissions-intensive goods and services, increasing efficiency gains, increasing use and development of low-carbon technologies, and reducing fossil fuel emissions.

The energy policy of the European Union has set a target of limiting the global temperature rise to 2 °C (3.6 °F) compared to preindustrial levels, of which 0.8 °C has already taken place and another 0.5–0.7 °C is already committed. The 2 °C rise is typically associated in climate models with a carbon dioxide equivalent concentration of 400–500 ppm by volume; the current (April 2011) level of carbon dioxide alone is 393 ppm by volume, and rising at 1-3 ppm annually. Hence, to avoid a very likely breach of the 2 °C target, CO_2 levels would have to be stabilised very soon; this is generally regarded as unlikely, based on current programmes in place to date. The importance of change is illustrated by the fact that world economic energy efficiency is presently improving at only half the rate of world economic growth.

Greenhouse Gas Concentrations and Stabilization

One of the issues often discussed in relation to climate change mitigation is the stabilization of greenhouse gas concentrations in the atmosphere. The United Nations Framework Convention on Climate Change (UNFCCC) has the ultimate objective of preventing "dangerous" anthropogenic (i.e., human) interference of the climate system. As is stated in Article 2 of the Convention, this requires that greenhouse gas (GHG) concentrations are stabilized in the atmosphere at a level where ecosystems can adapt naturally to climate change, food production is not threatened, and economic development can proceed in a sustainable fashion.

A distinction needs to be made between stabilizing GHG emissions and GHG concentrations. The two are not the same. The most important GHG emitted by human activities is carbon dioxide (chemical formula: CO_2). Stabilizing emissions of CO_2 at current levels would not lead to a stabilization in the atmospheric concentration of CO_2. In fact, stabilizing emissions at current levels would result in the atmospheric concentration of CO_2 continuing to rise over the 21st century and beyond.

The reason for this is that human activities are adding CO_2 to the atmosphere far faster than natural processes can remove it. This is analogous to a flow of water into a bathtub. So long as the tap runs water (analogous to the emission of carbon dioxide) into the tub faster than water escapes through the plughole (the natural removal of carbon dioxide from the atmosphere), then the level of water in the tub (analogous to the concentration of carbon dioxide in the atmosphere) will continue to rise.

Stabilizing the atmospheric concentration of the other greenhouse gases humans emit also depends on how fast their emissions are added to the atmosphere, and how fast the GHGs are removed. Stabilization for these gases is described in the later section on non-CO_2 GHGs.

Methods and Means

At the core of most proposals is the reduction of greenhouse gas emissions through reducing energy waste and switching to cleaner energy sources. Frequently discussed energy conservation methods include increasing the fuel efficiency of vehicles (often through hybrid, plug-in hybrid, and electric cars and improving conventional automobiles), individual-lifestyle changes and changing business practices. Newly developed technologies and currently available technologies including renewable energy (such as solar power, tidal and ocean energy, geothermal power, and wind power) and more controversially nuclear power and the use of carbon sinks, carbon credits, and taxation are aimed more precisely at countering continued greenhouse gas emissions. The ever-increasing global population and the planned growth of national GDPs based on current technologies are counter-productive to most of these proposals.

Alternative Energy Sources

Renewable Energy

Climate change concerns and the need to reduce carbon emissions are driving increasing growth in the renewable energy industries. Some 85 countries now have targets for their own renewable energy futures, and have enacted wide-ranging public policies to promote renewables. Low-carbon renewable energy replaces conventional fossil fuels in three main areas: power generation, hot water/ space heating, and transport fuels. Scientists have advanced a plan to power 100% of the world's energy with wind, hydroelectric, and solar power by the year 2030.

In terms of power generation, renewable energy currently provides 18 percent of total electricity generation worldwide and this percentage is growing each year. Renewable power generators are spread across many countries, and wind power alone already provides a significant share of electricity in some areas: for example, 14 percent in the U.S. state of Iowa, 40 percent in the northern German state of Schleswig-Holstein, and 20 percent in Denmark. Some countries get most of their power from renewables, including Iceland (100 percent), Brazil (85 percent), Austria (62 percent), New Zealand (65 percent), and Sweden (54 percent). Solar water heating makes an important and growing contribution in many countries, most notably in China, which now has 70 percent of the global total (180 GWth).

Worldwide, total installed solar water heating systems meet a portion of the water heating needs of over 70 million households. The use of biomass for heating continues to grow as well. In Sweden, national use of biomass energy has surpassed that of oil. Direct geothermal heating is also growing rapidly. Renewable biofuels for transportation, such as ethanol fuel and biodiesel, have contributed to a significant decline in oil consumption in the United States since 2006. The 93 billion liters of biofuels produced worldwide in 2009 displaced the equivalent of an estimated 68 billion liters of gasoline, equal to about 5 percent of world gasoline production.

Nuclear Power

Nuclear power currently produces 13-14% of the world's electricity. Since about 2001 the term nuclear renaissance has been used to refer to a possible nuclear power industry revival, driven by rising fossil fuel prices and new concerns about meeting greenhouse gas emission limits. At the same time, various barriers to a nuclear renaissance have been identified. These barriers include unfavourable economics compared to other sources of energy and slowness in addressing climate change.

New reactors under construction in Finland and France, which were meant to lead a nuclear renaissance, have been delayed and are running over-budget. China has 20 new reactors under construction, and there are also a considerable number of new reactors being built in South Korea, India, and Russia. At least 100 older and smaller reactors will "most probably be closed over the next 10-15 years". Nuclear power brings with it important waste disposal, safety, and security risks which are unique among low-carbon energy sources. Public attitudes towards nuclear power remain ambiguous in many developed countries, with significant anti-nuclear opposition even when majority opinion is in favour.

Carbon Intensity of Fossil Fuels

Natural gas (predominantly methane) produces less greenhouses gases per energy unit gained than oil which in turn produces less than coal, principally because coal has a larger ratio of carbon to hydrogen. The combustion of natural gas emits almost 30 percent less carbon dioxide than oil, and just under 45 percent less carbon dioxide than coal. In addition, there are also other environmental benefits.

A study performed by the Environmental Protection Agency (EPA) and the Gas Research Institute (GRI) in 1997 sought to discover whether the reduction in carbon dioxide emissions from increased natural gas (predominantly methane) use would be offset by a possible increased level of methane emissions from sources such as leaks and emissions. The study concluded that the reduction in emissions from increased natural gas use strongly outweighs the detrimental effects of increased methane emissions. Thus the increased use of natural gas in the place of other, dirtier fossil fuels can serve to lessen the emission of greenhouse gases in the United States. Most mitigation proposals imply — rather than directly state — an eventual reduction in global fossil fuel production. Also proposed are direct quotas on global fossil fuel production.

Energy Efficiency and Conservation

Efficient energy use, sometimes simply called "energy efficiency", is the goal of efforts to reduce the amount of energy required to provide products and services. For example, insulating a home allows a building to use less heating and cooling energy to achieve and maintain a comfortable temperature. Installing fluorescent lights or natural skylights reduces the amount of energy required to attain the same level of illumination compared to using traditional incandescent light bulbs. Compact fluorescent lights use two-thirds less energy and may last 6 to 10 times longer than incandescent lights.

Energy efficiency has proved to be a cost-effective strategy for building economies without necessarily growing energy consumption. For example, the state of California began implementing energy-efficiency measures in the mid-1970s, including building code and appliance standards with strict efficiency requirements. During the following years, California's energy consumption has remained approximately flat on a per capita basis while national U.S. consumption doubled. As part of its strategy, California implemented a "loading order" for new energy resources that puts energy efficiency first, renewable electricity supplies second, and new fossil-fired power plants last.

Energy conservation is broader than energy efficiency in that it encompasses using less energy to achieve a lesser energy service, for example through behavioural change, as well as encompassing energy efficiency. Examples of conservation without efficiency improvements would be heating a room less in winter, driving less, or working in a less brightly lit room. As with other definitions, the boundary between efficient energy use and energy conservation can be fuzzy, but both are important in environmental and economic terms. This is especially the case when actions are directed at the saving of fossil fuels. Reducing energy use is seen as a key solution to the problem of reducing greenhouse gas emissions. According to the International Energy Agency, improved energy efficiency in buildings, industrial processes and transportation could reduce the world's energy needs in 2050 by one third, and help control global emissions of greenhouse gases.

Transport

Modern energy efficient technologies, such as plug-in hybrid electric vehicles, and development of new technologies, such as hydrogen cars, may reduce the consumption of petroleum and emissions of carbon dioxide. A shift from air transport and truck transport to electricrail transport would reduce emissions significantly.

Increased use of biofuels (such as ethanol fuel and biodiesel that can be used in today's diesel and gasoline engines) could also reduce emissions if produced environmentally efficiently, especially in conjunction with regular hybrids and plug-in hybrids. For electric vehicles, the reduction of carbon emissions will improve further if the way the required electricity is generated is low-carbon (from renewable energysources). Effective urban planning to reduce sprawl would decrease Vehicle Miles Travelled (VMT), lowering emissions from transportation. Increased use of public transport can also reduce greenhouse gas emissions per passenger kilometre.

Urban Planning

Urban planning also has an effect on energy use. Between 1982 and 1997, the amount of land consumed for urban development in the United States increased by 47 percent while the nation's population grew by only 17 percent. Inefficient land use development practices have increased infrastructure costs as well as the amount of energy needed for transportation, community services, and buildings. At the same time, a growing number of citizens and government officials have begun advocating a smarter approach to land use planning. These smart growth practices include compact community development, multiple transportation choices, mixed land uses, and practices to conserve green space. These programmes offer environmental, economic, and quality-of-life benefits; and they also serve to reduce energy usage and greenhouse gas emissions.

Approaches such as New Urbanism and Transit-oriented development seek to reduce distances travelled, especially by private vehicles, encourage public transit and make walking and cycling more attractive options. This is achieved through medium-density, mixed-use planningand the concentration of housing within walking distance of town centres and transport nodes.

Smarter growth land use policies have both a direct and indirect effect on energy consuming behaviour. For example, transportation energy usage, the number one user of petroleum fuels, could be significantly reduced through more compact and mixed use land development patterns, which in turn could be served by a greater variety of non-automotive based transportation choices.

Building Design

Emissions from housing are substantial, and government-supported energy efficiency programmes can make a difference. For institutions of higher learning in the United States, greenhouse gas emissions depend primarily on total area of buildings and secondarily on climate. If climate is not taken into account, annual greenhouse gas emissions due to energy consumed on campuses plus purchased electricity can be estimated with the formula, $E=aS^b$, where a =0.001621 metric tonnes of CO_2 equivalent/square foot or 0.0241 metric tonnes of CO_2 equivalent/ square meter and b = 1.1354.

New buildings can be constructed using passive solar building design, low-energy building, or zero-energy building techniques, using renewable heat sources. Existing buildings can be made more efficient through the use of insulation, high-efficiency appliances (particularly hot water heaters and furnaces), double- or triple-glazed gas-filled windows, external window shades, and building orientation and siting.

Renewable heat sources such as shallow geothermal and passive solar energy reduce the amount of greenhouse gasses emitted. In addition to designing buildings which are more energy efficient to heat, it is possible to design buildings that are more energy efficient to cool by using lighter-coloured, more reflective materials in the development of urban areas (e.g. by painting roofs white) and planting trees. This saves energy because it cools buildings and reduces the urban heat island effect thus reducing the use of air conditioning.

Reforestation and Avoided Deforestation

Almost 20% (8 $GtCO_2$/year) of total greenhouse-gas emissions were from deforestation in 2007. The Stern Review found that, based on theopportunity costs of the landuse that would no longer be available for agriculture if deforestation were avoided, emission savings from avoided deforestation could potentially reduce CO_2 emissions for under \$5/tCO_2, possiblly as little as \$1/tCO_2. Afforestation and reforestation could save at least another 1$GtCO_2$/year, at an estimated cost of \$5/ tCO_2 to \$15/tCO_2. The Review determined these figures by assessing 8 countries responsible for 70% of global deforestation emissions.

Pristine temperate forest has been shown to store three times more carbon than IPCC estimates took into account, and 60% more carbon than plantation forest. Preventing these forests from being logged would have significant effects.

Further significant savings from other non-energy-related-emissions could be gained through cuts to agricultural emissions, fugitive emissions, waste emissions, and emissions from various industrial processes.

Eliminating waste Methane

Methane is a significantly more powerful greenhouse gas than carbon dioxide. Burning one molecule of methane generates one molecule of carbon dioxide. Accordingly, burning methane which would otherwise be released into the atmosphere (such as at oil wells, landfills, coal mines, waste treatment plants, etc.) provides a net greenhouse gas emissions benefit. However, reducing the amount of waste methane produced in the first place has an even greater beneficial impact, as might other approaches to productive use of otherwise-wasted methane.

In terms of prevention, vaccines are in the works in Australia to reduce significant global warming contributions from methane released by livestock via flatulence and eructation.

Geoengineering

Geoengineering is seen by some as an alternative to mitigation and adaptation, but by others as an entirely separate response to climate change. In a literature assessment, Barker *et al.* (2007) described geoengineering as a type of mitigation policy. IPCC (2007) concluded that geoengineering options, such as ocean fertilization to remove CO_2 from the atmosphere, remained largely unproven. It was judged that reliable cost estimates for geoengineering had not yet been published.

Chapter 28 of the National Academy of Sciences report *Policy Implications of Greenhouse Warming: Mitigation, Adaptation, and the Science Base* (1992) defined geoengineering as "options that would involve large-scale engineering of our environment in order to combat or counteract the effects of changes in atmospheric chemistry." They evaluated a range of options to try to give preliminary answers to two questions: can these options work and could they be carried out with a reasonable cost. They also sought to encourage discussion of a third question — what adverse side effects might there be. The following types of option were examined: reforestation, increasing ocean absorption of carbon dioxide (carbon sequestration) and screening out some sunlight. NAS also argued "Engineered countermeasures need to be evaluated but should not be implemented without broad understanding of the direct effects and the potential side effects, the ethical issues, and the risks.".

Greenhouse Gas Remediation

Carbon sequestration has been proposed as a method of reducing the amount of radiative forcing. Carbon sequestration is a term that describes processes that remove carbon from the atmosphere. A variety of means of artificially capturing and storing carbon, as well as of enhancing natural sequestration processes, are being explored. The main natural process is photosynthesis by plants and single-celled organisms. Artificial processes vary, and concerns have been expressed about their long-term effects.

Although they require land, natural sinks can be enhanced by reforestation and afforestation carbon offsets, which fix carbon dioxide for as little as $0.11 per metric ton.

Sulfate Aerosols

The ability of stratospheric sulfate aerosols to create a global dimming effect has made them a possible candidate for use in geoengineering projects.

Biomass Energy

During its growth, vegetation traps carbon dioxide from the atmosphere through photosynthesis. When this biomass decomposes or is combusted, the carbon is again released as carbon dioxide. This process is part of the global carbon cycle. Through the use of biomass for energy and materials, e.g. in biomass fuelled power plants, parts of this cycle is controlled by man. However, whether direct use of biomass for energy can be carbon neutral is case-specific and remains a matter of controversy.

Combining a biomass energy system with carbon capture and storage technology (a form of Geoengineering, is so-called bio-energy with carbon capture and storage (BECCS). Proponents of BECCS, a technology yet to be proven, hope that it will result in net-negative carbon dioxide emissions, i.e. net removal of carbon dioxide from the atmosphere. In comparison with other geoengineering options, BECCS has been suggested as a low-risk, near-term tool to effectively remove carbon from the atmosphere. Even so, whether biomass can be sustainably obtained in significant quantities remains controversial. In July 2011 a report by the United States Government Accountability Office on geoengineering found that "climate engineering technologies do not now offer a viable response to global climate change."

Carbon air Capture

It is notable that the availability of cheap energy and appropriate sites for geological storage of carbon may make carbon dioxide air capture viable commercially. It is, however, generally expected that carbon dioxide air capture may be uneconomic when compared to carbon capture and storage from major sources — in particular, fossil fuel powered power stations, refineries, etc. In such cases, costs of energy produced will grow significantly. However, captured CO_2 can be used to force more crude oil out of oil fields, as Statoil andShell have made plans to do. CO_2 can also be used in commercial greenhouses, giving an opportunity to kick-start the technology. Some attempts have been made to use algae to capture smokestack emissions, notably the Green Fuel Technologies Corporation, who have now shut down operations.

Carbon Capture and Storage

Carbon capture and storage (CCS) is a plan to mitigate climate change by capturing carbon dioxide (CO_2) from large point sources such as power plants and subsequently storing it away safely instead of releasing it into the atmosphere. The Intergovernmental Panel on Climate Change says CCS could contribute between 10% and 55% of the cumulative worldwide carbon-mitigation effort over the next 90 years. The Agency says CCS is "the most important single new technology for CO_2 savings" in power generation and industry.

Though it requires up to 40% more energy to run a CCS coal power plant than a regular coal plant, CCS could potentially capture about 90% of all the carbon emitted by the plant. Norway, which first began storing CO_2, has cut its emissions by almost a million tons a year, or about 3% of the country's 1990 levels. Please see also direct conversion of CO_2 to fuels. As of late 2011, the total CO2 storage capacity of all 14 projects in operation or under construction is over 33 million tonnes a year. This is broadly equivalent to preventing the emissions from more than six million cars from entering the atmosphere each year.

Pacala and Socolow: 15 Programmes

Pacala and Socolow of Princeton have proposed a programme to reduce CO_2 emissions by 1 billion metric tons per year " or 25 billion tons over the 50-year period. The proposed 15 different programmes, any seven of which could achieve the goal, are:

1. more efficient vehicles " increase fuel economy from 30 to 60 mpg (7.8 to 3.9 L/100 km) for 2 billion vehicles,
2. reduce use of vehicles " improve urban design to reduce miles driven from 10,000 to 5,000 miles (16,000 to 8,000 km) per year for 2 billion vehicles,
3. efficient buildings " reduce energy consumption by 25%,
4. improve efficiency of coal plants from today's 40% to 60%,
5. replace 1,400 GW (gigawatt) of coal power plants with natural gas,

6. capture and store carbon emitted from 800 GW of new coal plants,
7. capture and reuse hydrogen created by No. 6 above,
8. capture and store carbon from coal to syn fuels conversion at 30 million barrels per day (4,800,000 m^3/d),
9. displace 700 GW of coal power with nuclear,
10. add 2 million 1 MW wind turbines (50 times current capacity),
11. displace 700 GW of coal with 2,000 GW (peak) solar power (700 times current capacity),
12. produce hydrogen fuel from 4 million 1 MW wind turbines,
13. use biomass to make fuel to displace oil (100 times current capacity),
14. stop de-forestation and re-establish 300 million hectares of new tree plantations,
15. conservation tillage " apply to all crop land (10 times current usage).

Nature.com argued in June 2008 that "If we are to have confidence in our ability to stabilize carbon dioxide levels below 450 p.p.m. emissions must average less than 5 billion metric tons of carbon per year over the century. This means accelerating the deployment of the wedges so they begin to take effect in 2015 and are completely operational in much less time than originally modelled by Socolow and Pacala."

Societal Controls

Another method being examined is to make carbon a new currency by introducing tradeable "Personal Carbon Credits". The idea being it will encourage and motivate individuals to reduce their 'carbon footprint' by the way they live. Each citizen will receive a free annual quota of carbon that they can use to travel, buy food, and go about their business. It has been suggested that by using this concept it could actually solve two problems; pollution and poverty, old age pensioners will actually be better off because they fly less often, so they can cash in their quota at the end of the year to pay heating bills, etc.

Population

Various organizations promote population control as a means for mitigating global warming. Proposed measures include improving access to family planning and reproductive health care and information, reducing natalistic politics, public education about the consequences of continued population growth, and improving access of women to education and economic opportunities. Population control efforts are impeded by there being somewhat of a taboo in some countries against considering any such efforts. Also, various religions discourage or prohibit some or all forms of birth control.

Population size has a different per capita effect on global warming in different countries, since the per capita production of anthropogenic greenhouse gases varies greatly by country.

Non-CO_2 Greenhouse Gases

CO_2 is not the only GHG relevant to mitigation, and governments have acted to regulate the emissions of other GHGs emitted by human activities (anthropogenic GHGs). The emissions caps agreed to by most developed countries under the Kyoto Protocol regulate the emissions of almost all the anthropogenic GHGs.

These gases are CO_2, methane (chemical formula: CH_4), nitrous oxide(N_2O), the hydrofluorocarbons (abbreviated HFCs), perfluorocarbons (PFCs), and sulfur hexafluoride (SF_6). Stabilizing the atmospheric concentrations of the different anthropogenic GHGs requires an understanding of their different physical properties. Stabilization depends both on how quickly GHGs are added to the atmosphere and how fast they are removed. The rate of removal is measured by the atmospheric lifetime of the GHG in question.

Here, the lifetime is defined as the time required for a given perturbation of the GHG in the atmosphere to be reduced to 37% of its initial amount. Methane has a relatively short atmospheric lifetime of about 12 years, while N_2O's lifetime is about 110 years. For methane, a reduction of about 30% below current emission levels would lead to a stabilization in its atmospheric concentration, while for N_2O, an emissions reduction of more than 50% would be required.

Another physical property of the anthropogenic GHGs relevant to mitigation is the different abilities of the gases to trap heat (in the form of infrared radiation). Some gases are more effective at trapping heat than others, e.g., SF_6 is 22,200 times more effective a GHG than CO_2 on a per-kilogram basis.

A measure for this physical property is the global warming potential (GWP), and is used in the Kyoto Protocol. Although not designed for this purpose, the Montreal Protocol has probably benefitted climate change mitigation efforts. The Montreal Protocol is an international treaty that has successfully reduced emissions of ozone-depleting substances (e.g., CFCs), which are also greenhouse gases.

Costs and Benefits

Costs

The Stern Review proposes stabilising the concentration of greenhouse-gas emissions in the atmosphere at a maximum of 550ppm CO_2e by 2050. The Review estimates that this would mean cutting total greenhouse-gas emissions to three quarters of 2007 levels. The Review further estimates that the cost of these cuts would be in the range –1.0 to +3.5% of WorldGDP, (i.e. GWP), with an average estimate of approximately 1%. Stern has since revised his estimate to 2% of GWP. For comparison, the Gross World Product (GWP) at PPPwas estimated at $74.5 trillion in 2010, thus 2% is approximately $1.5 trillion. The Review emphasises that these costs are contingent on steady reductions in the cost of low-carbon technologies. Mitigation costs will also vary according to how and when emissions are cut: early, well-planned action will minimise the costs. One way of estimating the cost of reducing emissions is by considering the likely costs of potential technological and output changes. Policy makers can compare the marginal abatement costs of different methods to assess the cost and amount of possible abatement over time. The marginal abatement costs of the various measures will differ by country, by sector, and over time.

Benefits

Yohe *et al.* (2007) assessed the literature on sustainability and climate change. With high confidence, they suggested that up to the year 2050, an effort to cap greenhouse gas (GHG) emissions at 550 ppm would benefit developing countries significantly. This was judged to be especially the case when combined with enhanced adaptation. By 2100, however, it was still judged likely that there would be significant climate change impacts. This was judged to be the case even with aggressive mitigation and significantly enhanced adaptive capacity.

Sharing

One of the aspects of mitigation is how to share the costs and benefits of mitigation policies. There is no scientific consensus over how to share these costs and benefits (Toth *et al.*, 2001). In terms of the politics of mitigation, the UNFCCC's ultimate objective is to stabilize concentrations of GHG in the atmosphere at a level that would prevent "dangerous" climate change (Rogner *et al.*, 2007). There is, however, no widespread agreement on how to define "dangerous" climate change.

GHG emissions are an important correlate of wealth, at least at present (Banuri *et al.*, 1996, pp. 91–92). Wealth, as measured by per capita income (i.e., income per head of population), varies widely between different countries. Activities of the poor that involve emissions of GHGs are often associated with basic needs, such as heating to stay tolerably warm. In richer countries, emissions tend to be associated with things like cars, central heating, etc. The impacts of cutting emissions could therefore have different impacts on human welfare according wealth.

Distributing Emissions Abatement Costs

There have been different proposals on how to allocate responsibility for cutting emissions (Banuri *et al.*, 1996, pp. 103–105):

- *Egalitarianism:* this system interprets the problem as one where each person has equal rights to a global resource, i.e., polluting the atmosphere.
- *Basic needs and Rawlsian criteria:* this system would have emissions allocated according to basic needs, as defined according to a minimum level of consumption. Consumption above basic needs would require countries to buy more emission rights. This can be related to Rawlsian philosophy. From this viewpoint, developing countries would need to be at least as well off under an emissions control regime as they would be outside the regime.
- *Proportionality and polluter-pays principle:* Proportionality reflects the ancient Aristotelian principle that people should receive in proportion to what they put in, and pay in proportion to the damages they cause. This has a potential relationship with the "polluter-pays principle", which can be interpreted in a number of ways:
- *Historical responsibilities*: this asserts that allocation of emission rights should be based on patterns of past emissions. Two-thirds of the stock of GHGs in the atmosphere at present is due to the past actions of developed countries (Goldemberg *et al.*, 1996, p. 29).
- *Comparable burdens and ability to pay*: with this approach, countries would reduce emissions based on comparable burdens and their ability to take on the costs of reduction. Ways to assess burdens include monetary costs per head of population, as well as other, more complex measures, like the UNDP's Human Development Index.
- *Willingness to pay*: with this approach, countries take on emission reductions based on their ability to pay along with how much they benefit from reducing their emissions.

Specific Proposals

- *Ad hoc:* Lashof (1992) and Cline (1992) (referred to by Banuri *et al.*, 1996, p. 106), for example, suggested that allocations based partly on GNP could be a way of sharing the burdens of emission reductions. This is because GNP and economic activity are partially tied to carbon emissions.
- *Equal per capita entitlements:* this is the most widely cited method of distributing abatement costs, and is derived from egalitarianism (Banuri *et al.*, 1996, pp. 106–107). This approach can be divided into two categories. In the first category, emissions are allocated according to national population. In the second category, emissions are allocated in a way that attempts to account for historical (cumulative) emissions.
- *Status quo:* with this approach, historical emissions are ignored, and current emission levels are taken as a status quo right to emit (Banuri *et al.*, 1996, p. 107). An analogy for this approach can be made with fisheries, which is a common, limited resource. The analogy would be with the atmosphere, which can be viewed as an exhaustible natural resource(Goldemberg *et al.*, 1996, p. 27). In international law, one state recognized the long-established use of another state's use of the fisheries resource. It was also recognized by the state that part of the other state's economy was dependent on that resource.

Governmental and Intergovernmental Action

Many countries, both developing and developed, are aiming to use cleaner technologies (World Bank, 2010, p. 192). Use of these technologies aids mitigation and could result in substantial reductions in CO_2 emissions. Policies include targets for emissions reductions, increased use of renewable energy, and increased energy efficiency. It is often argued that the results of climate change are more damaging in poor nations, where infrastructures are weak and few social services exist. The Commitment to Development Index is one attempt to analyze rich country policies taken to reduce their disproportionate use of the global commons. Countries do well if their greenhouse gas emissions are falling, if their gas taxes are high, if they do not subsidize the fishing industry, if they have a low fossil fuel rate per capita, and if they control imports of illegally cut tropical timber.

Kyoto Protocol

The main current international agreement on combating climate change is the Kyoto Protocol, which came into force on 16 February

2005. The Kyoto Protocol is an amendment to theUnited Nations Framework Convention on Climate Change (UNFCCC). Countries that have ratified this protocol have committed to reduce their emissions of carbon dioxide and five other greenhouse gases, or engage in emissions trading if they maintain or increase emissions of these gases.

Copenhagen 2009

The first phase of the Kyoto Protocol expires in 2012. The United Nations Climate Change Conference in Copenhagen in December 2009 was the next in an annual series of UN meetings that followed the 1992 Earth Summit in Rio. In 1997 the talks led to the Kyoto Protocol, Copenhagen was considered the world's chance to agree a successor to Kyoto that would bring about meaningful carbon cuts.

Encouraging use Changes

Subsidies

A programme of subsidization balanced against expected flood costs could pay for conversion to 100% renewable power by 2030. The proponents of such a plan expect the cost to generate and transmit power in 2020 will be less than 4 cents per kilowatt hour (in 2007 dollars) for wind, about 4 cents for wave and hydroelectric, from 4 to 7 cents for geothermal, and 8 cents per kwh for solar, fossil, and nuclear power.

Carbon Emissions Trading

With the creation of a market for trading carbon dioxide emissions within the Kyoto Protocol, it is likely that London financial markets will be the centre for this potentially highly lucrative business; the New York and Chicago stock markets may have a lower trade volume than expected as long as the US maintains its rejection of the Kyoto.

However, emissions trading may delay the phase-out of fossil fuels. The European Union Emission Trading Scheme (EU ETS) is the largest multi-national, greenhouse gas emissions trading scheme in the world. It commenced operation on 1 January 2005, and all 25 member states of the European Union participate in the scheme which has created a new market in carbon dioxide allowances estimated at 35 billion Euros (US$43 billion) per year.

The Chicago Climate Exchange was the first (voluntary) emissions market, and is soon to be followed by Asia's first market (Asia Carbon Exchange). A total of 107 million metric tonnes of carbon dioxide equivalent have been exchanged through projects in 2004, a 38%

increase relative to 2003 (78 Mt CO_2e). Twenty three multinational corporations have come together in the G8 Climate Change Roundtable, a business group formed at the January 2005 World Economic Forum. The group includes Ford, Toyota, British Airways and BP. On 9 June 2005 the Group published a statement stating that there was a need to act on climate change and claiming that market-based solutions can help. It called on governments to establish "clear, transparent, and consistent price signals" through "creation of a long-term policy framework" that would include all major producers of greenhouse gases.

The Regional Greenhouse Gas Initiative is a proposed carbon trading scheme being created by nine North-eastern and Mid-Atlantic American states; Connecticut, Delaware, Maine, Massachusetts, New Hampshire, New Jersey, New York, Rhode Island and Vermont. The scheme was due to be developed by April 2005 but has not yet been completed.

Emissions Tax

An emissions tax on greenhouse gas emissions requires individual emitters to pay a fee, charge or tax for every tonne of greenhouse gas released into the atmosphere. Most environmentally related taxes with implications for greenhouse gas emissions in OECD countries are levied on energy products and motor vehicles, rather than on CO_2 emissions directly.

Emission taxes can be both cost effective and environmentally effective. Difficulties with emission taxes include their potential unpopularity, and the fact that they cannot guarantee a particular level of emissions reduction. Emissions or energy taxes also often fall disproportionately on lower income classes. In developing countries, institutions may be insufficiently developed for the collection of emissions fees from a wide variety of sources.

Implementation

Implementation puts into effect climate change mitigation strategies and targets. These can be targets set by international bodies or voluntary action by individuals or institutions. This is the most important, expensive and least appealing aspect of environmental governance.

Funding

Implementation requires funding sources but is often beset by disputes over who should provide funds and under what conditions. A lack of funding can be a barrier to successful strategies as there are no formal arrangements to finance climate change development and implementation. Funding is often provided by nations, groups of nations and increasingly NGO and private sources. These funds are often channelled through the Global Environmental Facility (GEF). This is an environmental funding mechanism in the World Bank which is designed to deal with global environmental issues.

The GEF was originally designed to tackle four main areas: biological diversity, climate change, international waters and ozone layer depletion, to which land degradation and persistent organic pollutant were added. The GEF funds projects that are agreed to achieve global environmental benefits that are endorsed by governments and screened by one of the GEF's implementing agencies.

Problems

There are numerous issues which result in a current perceived lack of implementation. It has been suggested that the main barriers to implementation are, Uncertainty, Fragmentation, Institutional void, Short time horizon of policies and politicians and Missing motives and willingness to start adapting. The relationships between many climatic processes can cause large levels of uncertainty as they are not fully understood and can be a barrier to implementation. When information on climate change is held between the large numbers of actors involved it can be highly dispersed, context specific or difficult to access causing fragmentation to be a barrier.

Institutional void is the lack of commonly accepted rules and norms for policy processes to take place, calling into question the legitimacy and efficacy of policy processes. The Short time horizon of policies and politicians often means that climate change policies are not implemented in favour of socially favoured societal issues. Statements are often posed to keep the illusion of political action to prevent or postpone decisions being made. Missing motives and willingness to start adapting is a large barrier as it prevents any implementation.

Occurrence

Despite a perceived lack of occurrence, evidence of implementation is emerging internationally. Some examples of this are the initiation of NAPA's and of joint implementation. Many developing nations have made National Adaptation Programmes of Action (NAPAs) which are frameworks to prioritize adaption needs. The implementation of many of these is supported by GEF agencies. Many developed countries are implementing 'first generation' institutional adaption plans particularly at the state and local government scale. There has also been a push towards joint implementation between countries by the UNFCC as this has been suggested as a cost effective way for objectives to be achieved.

Territorial Policies

United States

Efforts to reduce greenhouse gas emissions by the United States include energy policies which encourage efficiency through programmes like Energy Star, Commercial Building Integration, and the Industrial Technologies Programme. On 12 November 1998, Vice President Al Gore symbolically signed the Kyoto Protocol, but he indicated participation by the developing nations was necessary prior its being submitted for ratification by the United States Senate. In 2007, Transportation Secretary Mary Peters, with White House approval, urged governors and dozens of members of the House of Representatives to block California's first-in-the-nation limits on greenhouse gases from cars and trucks, according to e-mails obtained by Congress. The U.S. Climate Change Science Programme is a group of about twenty federal agencies and US Cabinet Departments, all working together to address global warming.

The Bush administration pressured American scientists to suppress discussion of global warming, according to the testimony of the Union of Concerned Scientists to the Oversight and Government Reform Committee of the U.S. House of Representatives. "High-quality science" was "struggling to get out," as the Bush administration pressured scientists to tailor their writings on global warming to fit the Bush administration's skepticism, in some cases at the behest of an ex-oil industry lobbyist.

"Nearly half of all respondents perceived or personally experienced pressure to eliminate the words 'climate change,' 'global warming' or other similar terms from a variety of communications." Similarly, according to the testimony of senior officers of the Government Accountability Project, the White House attempted to bury the report "National Assessment of the Potential Consequences of Climate Variability and Change," produced by U.S. scientists pursuant to U.S. law. Some U.S. scientists resigned their jobs rather than give in to White House pressure to underreport global warming.

In the absence of substantial federal action, state governments have adopted emissions-control laws such as the Regional Greenhouse Gas Initiative in the Northeast and the Global Warming Solutions Act of 2006 in California.

Developing Countries

In order to reconcile economic development with mitigating carbon emissions, developing countries need particular support, both financial and technical. One of the means of achieving this is the Kyoto Protocol's Clean Development Mechanism (CDM). The World Bank's Prototype Carbon Fund is a public private partnership that operates within the CDM.

An important point of contention, however, is how overseas development assistance not directly related to climate change mitigation is affected by funds provided to climate change mitigation. One of the outcomes of the UNFCC Copenhagen Climate Conference was the Copenhagen Accord, in which developed countries promised to provide US $30 million between 2010–2012 of new and additional resources. Yet it remains unclear what exactly the definition of additional is and the European Commission has requested its member states to define what they understand to be additional, and researchers at the Overseas Development Institute have found 4 main understandings:

1. Climate finance classified as aid, but additional to (over and above) the '0.7%' ODA target;
2. Increase on previous year's Official Development Assistance (ODA) spent on climate change mitigation;
3. Rising ODA levels that include climate change finance but where it is limited to a specified percentage; and
4. Increase in climate finance not connected to ODA.

The main point being that there is a conflict between the OECD states budget deficit cuts, the need to help developing countries adapt to develop sustainably and the need to ensure that funding does not come from cutting aid to other important Millennium Development Goals.

In July 2005 the U.S., China, India, Australia, as well as Japan and South Korea, agreed to the Asia-Pacific Partnership for Clean Development and Climate. The pact aims to encourage technological development that may mitigate global warming, without coordinated emissions targets.

The highest goal of the pact is to find and promote new technology that aid both growth and a cleaner environment simultaneously. An example is the Methane to Markets initiative which reduces methane emissions into the atmosphere by capturing the gas and using it for growth enhancing clean energy generation. Critics have raised concerns that the pact undermines the Kyoto Protocol.

However, none of these initiatives suggest a quantitative cap on the emissions from developing countries. This is considered as a particularly difficult policy proposal as the economic growth of developing countries are proportionally reflected in the growth of greenhouse emissions. Critics of mitigation often argue that, the developing countries' drive to attain a comparable living standard to the developed countries would doom the attempt at mitigation of global warming. Critics also argue that holding down emissions would shift the human cost of global warming from a general one to one that was borne most heavily by the poorest populations on the planet.

In an attempt to provide more opportunities for developing countries to adapt clean technologies, UNEP and WTO urged the international community to reduce trade barriers and to conclude the Doha trade round "which includes opening trade in environmental goods and services".

Non-governmental Approaches

While many of the proposed methods of mitigating global warming require governmental funding, legislation and regulatory action, individuals and businesses can also play a part in the mitigation effort.

Choices in Personal Actions and Business Operations

Environmental groups encourage individual action against global warming, often aimed at the consumer. Common recommendations include lowering home heating and cooling usage, burning less gasoline, supporting renewable energy sources, buying local products to reduce transportation, turning off unused devices, and various others.

A geophysicist at Utrecht University has urged similar institutions to hold the vanguard in voluntary mitigation, suggesting the use of communications technologies such asvideoconferencing to reduce their dependence on long-haul flights.

Air Travel and Shipment

Climate scientist Kevin Anderson raised concern about the growing effect of rapidly increasing global air transport on the climate in a paper and a presentation in 2008, suggesting that reversing this trend is necessary. Part of the difficulty is that when aviation emissions are made at high altitude, the climate impacts are much greater than otherwise. Others have been raising the related concerns of the increasing hypermobility of individuals, whether travelling for business or pleasure, involving frequent and often long distance air travel, as well as air shipment of goods.

Business Opportunities and Risks

On 9 May 2005 Jeff Immelt, the chief executive of General Electric (GE), announced plans to reduce GE's global warming related emissions by one percent by 2012. "GE said that given its projected growth, those emissions would have risen by 40 percent without such action."

On 21 June 2005 a group of leading airlines, airports and aerospace manufacturers pledged to work together to reduce the negative environmental impact of aviation, including limiting the impact of air travel on climate change by improving fuel efficiency and reducing carbon dioxide emissions of new aircraft by fifty percent per seat kilometre by 2020 from 2000 levels.

The group aims to develop a common reporting system for carbon dioxide emissions per aircraft by the end of 2005, and pressed for the early inclusion of aviation in the European Union's carbon emission trading scheme.

Bibliography

Alexander, W. H.: *Climatological History of Ohio*, Ohio State University Engineering Experiment Station, Columbus, OH, 1924.

Aubrey Meyer: *Contraction and Convergence: The Global Solution to Climate Change*, Schumacher Briefing 5, Green Books, Totnes, Devon, UK, 2005.

Bartok J.W.: *Lower Humidity Levels in Your Greenhouse,* Cooperative Extension System, University of Connecticut, Publication SEG, 1990.

Blair, T. A., and Fite, R. C.: *Weather Elements*, Prentice-Hall, Englewood Cliffs, New Jersey, 1965.

Carey, Mark: *In the Shadow of Melting Glaciers: Climate Change and Andean Society,* Oxford University Press, Oxford, 2010.

Cunningham, Anne S.: *Crystal Palaces: Garden Conservatories of the United States,* Princeton Architectural Press, New York, 2000.

David Lewis Feldman: *The Energy Crisis: Unresolved Issues and Enduring Legacies,* Johns Hopkins UP, London, 1996.

David Michel: *Climate Policy for the 21st Century,* Brookings Institution, London, 2004.

Ells, J. E., J. D. Butler, J. J. Hanan and W. D. Holley: *Hydroponics-growing Plants without Soil*, Cooperative Extension Service, Colorado State Univ., 1991.

Erik Nissen-Peterson : *Rainwater Catchment Systems*, UK: Intermediate Technology Publications, 1999.

Fagan, Brian M: *The Little Ice Age: How Climate Made History, 1300-1850*, Basic Books, Boulder, 2000.

Frenzel, Burkhard: *Climatic Trends and Anomalies in Europe 1675-1715*, Gustav Fischer Verlag, Stuttgart, 1994.

Gough, Kathleen: *Rural change in southeast India: 1950s to 1980s*, Oxford University Press, Delhi, 1989.

Greenough, Paul R.: *Prosperity and Misery in Modern Bengal. The Famine of 1943-1944*, Oxford University Press, Delhi, 1982.

Guha, Ramachandra: *The Unquiet Woods: Ecological Change and Peasant Resistance in the Himalaya,* Oxford University Press, Delhi, 1989.

Hall, David A.: *Role of Perlite in Hydroponic Culture*, South Federal, Chicago, 1991.

Helen Bannayan: *Water Resources of Jordan: Present Status and Future Potentials*, Amman, Friedrich-Ebert-Stiftung, 1993.

Issar, A. S.: *Water Shall Flow from the Rock* (Hydrogeology and Climate in the Lands of the Bible): Springer-Verlag, New York, 1990.

Jana B. L.: *Water Harvesting and Watershed Management*, Agrotech, Delhi, 2008.

Jha, Jagdish Chandra: *The Tribal Revolt in Chotanagpur, 1831-1832,* Kashi Prasad Jayaswal Research Institute, Patna, 1987.

Kaminsky, Arnold. P.: *The India Office, 1880-1910*, Greenwood Publishing Group, Connecticut, 1986.

Ladurie, Emmanuel le Roy: *Times of Feast, Times of Famine: A History of Climate Since the Year 1000,* Doubleday & Company, Inc., Garden City, NY, 1971.

Lamb, Hubert: *Climate, History, and the Modern World,* Routledge, London, 1995.

Manjunatha, S.R. : *An Introduction to Environmental Energy Resources*, Cyber Tech Publications, Delhi, 2010.

Oliver, John E.: *Encyclopedia of world climatology*, Springer, Delhi, 2005.

Pandian, M. S. S.: *The Political Economy of Agrarian Change: Nachilnadu, 1880-1939*, Sage Publications, New Delhi, 1990.

Panikkar, K. N.: *Against lord and state: religion and peasant uprisings in Malabar, 1836-1921*, Oxford University Press, Delhi, 1989.

Parrillo, Vincent N.: *Encyclopedia of social problems,* Sage Publications, Delhi, 2008.

Ray, Ratnalekha: *Change in Bengal Agrarian Society, c. 1760- 1850,* Manohar, New Delhi, 1979.

Richard W. Bulliet: *The Camel and the Wheel*, Harvard University Press, England, 1975.

Schwarz, M.: *Soilless Culture Management*, Springer-Verlag. Berlin, NY, 1995.

Singh, Harish C : *Energy Conservation : Tools, Techniques and Guidelines*, Swastik, Delhi, 2008.

White, Sam: *The Climate of Rebellion in the Early Modern Ottoman Empire*, Cambridge University Press, New York, 2011.

Index

❑❑❑